Corrosion of Steel in Concrete

'Rust's a Must'

Mighty ships upon the ocean
Suffer from severe corrosion,
Even those that stay at dockside
Are rapidly becoming oxide
Alas that piling in the sea
Is mostly Fe_2O_3,
And where the ocean meets the shore
You'll find there's Fe_3O_4,
'Cause when the wind is salt and gusty
Things are getting awful rusty.

We can measure we can test it,
We can halt it or arrest it,
We can gather it and weigh it,
We can coat it, we can spray it,
We examine and dissect it,
We cathodically protect it,
We can pick it up and drop it,
But heaven knows, we'll never stop it.

So here's to rust, no doubt about it,
Most of us would starve without it.

T.R.B. Watson, June 1974

Corrosion of Steel in Concrete
Understanding, investigation and repair

John P. Broomfield

Taylor & Francis
Taylor & Francis Group
LONDON AND NEW YORK

First published 1997 by Taylor & Francis

Reprinted 1998 by Taylor & Francis, an imprint of Routledge
2 Park Square, Milton Park, Abingdon, Oxon, OX14 4RN
270 Madison Ave, New York NY 10016

Transferred to Digital Printing 2005

© 1997 John P. Broomfield

Typeset in 10/12pt Palatino by Acorn Bookwork, Salisbury, Wiltshire

All rights reserved. No part of this book may be reprinted
or reproduced or utilized in any form or by any electronic,
mechanical, or other means, now known or hereafter
invented, including photocopying and recording, or in any
information storage or retrieval system, without permission
in writing from the publishers.

British Library Cataloguing in Publication Data
A catalogue record for this book is available from the British Library

Library of Congress Cataloguing in Publication Data
A catalogue record for this book is available from the Library of Congress

ISBN 0-419-19630-7

Printed and bound by Antony Rowe Ltd, Eastbourne

Contents

Preface	xiii
Acknowledgements	xv
Glossary	xvii

1 Introduction — 1

2 Corrosion of steel in concrete — 5
2.1 The corrosion process — 6
2.2 Black rust — 9
2.3 Pits, stray current and bacterial corrosion — 10
    2.3.1 Pit formation — 10
    2.3.2 Bacterial corrosion — 10
    2.3.3 Stray current induced corrosion — 12
    2.3.4 Local versus general corrosion (macrocells versus microcells) — 12
2.4 Electrochemistry, cells and half cells — 13
2.5 Conclusions — 15

3 Causes and mechanisms of corrosion and corrosion damage in concrete — 16
3.1 Carbonation — 16
    3.1.1 Carbonation transport through concrete — 19
3.2 Chloride attack — 20
    3.2.1 Sources of chlorides — 20
    3.2.2 Chloride transport through concrete — 22
    3.2.3 Chloride attack mechanism — 22
    3.2.4 Macrocell formation — 25
3.3 Corrosion damage — 25
3.4 Vertical cracks, horizontal cracks and corrosion — 28
3.5 The synergistic relationship between chloride and carbonation attack, chloride binding and release — 28
References — 28

4 Condition evaluation	30
4.1 Preliminary survey	31
4.2 Detailed survey	31
4.3 Available techniques	31
4.4 Visual inspection	33
4.4.1 Property to be measured	34
4.4.2 Equipment and use	34
4.4.3 Interpretation	34
4.4.4 Limitations	34
4.5 Delamination	36
4.5.1 Property to be measured	36
4.5.2 Equipment and use	36
4.5.3 Interpretation	37
4.5.4 Limitations	37
4.6 Cover	38
4.6.1 Property to be measured	39
4.6.2 Equipment and use	39
4.6.3 Interpretation	39
4.6.4 Limitations	39
4.7 Half cell potential measurements	40
4.7.1 Equipment and use	41
4.7.2 Interpretation and the ASTM criteria	44
4.7.3 Half cell potential mapping	48
4.7.4 Cell to cell potentials	48
4.8 Carbonation depth measurement	51
4.8.1 Equipment and use	51
4.8.2 Interpretation	51
4.8.3 Limitations	52
4.9 Chloride determination	53
4.9.1 Property to be measured	54
4.9.2 Equipment and use	55
4.9.3 Interpretation	57
4.10 Resistivity measurement	57
4.10.1 Property to be measured	60
4.10.2 Equipment and use	61
4.10.3 Interpretation	62
4.10.4 Limitations	62
4.11 Corrosion rate measurement	63
4.11.1 Property to be measured	64
4.11.2 Equipment and use: linear polarization	64
4.11.3 Carrying out a corrosion rate survey	67
4.11.4 Interpretation: linear polarization	68
4.11.5 Limitations: linear polarization	72
4.11.6 Equipment and use: macrocell techniques	73

4.11.7 Interpretation: macrocell techniques	73
4.12 Other useful tests for corrosion assessment	75
4.12.1 Permeability and absorption tests	75
4.12.2 Concrete characteristics: cement content, petrography, W/C ratio	76
4.12.3 Radar, radiography, PUNDIT, pulse velocity	76
4.13 Survey and assessment methodology	77
4.14 Monitoring	78
4.15 Special conditions: coated rebars and prestressing	79
4.15.1 Epoxy coated and galvanized reinforcing bars	79
4.15.2 Internal prestressing cables in ducts	80
References	81
5 Physical and chemical repair and rehabilitation techniques	**85**
5.1 Concrete removal and surface preparation	86
5.1.1 Pneumatic hammers	88
5.1.2 Hydrojetting	88
5.1.3 Milling machines	90
5.1.4 Comparative costing	92
5.1.5 Concrete damage and surface preparation	92
5.2 Patches	93
5.2.1 Incipient anodes	93
5.2.2 Load transfer and structural issues	96
5.3 Coatings, sealers, membranes and barriers	96
5.3.1 Carbonation repairs	97
5.3.2 Coatings against chlorides, penetrating sealers	98
5.3.3 Waterproofing membranes	99
5.3.4 Barriers and deflection systems	101
5.4 Encasement and overlays	102
5.5 Sprayed concrete	103
5.6 Corrosion inhibitors	104
References	106
6 Electrochemical repair techniques	**107**
6.1 Basic principles of electrochemical techniques	107
6.2 Cathodic protection	108
6.2.1 Theory and principles of impressed current systems	108
6.2.2 Theory and principles of sacrificial anode systems	111
6.2.3 The history of cathodic protection of steel in concrete	115
6.3 The components of an impressed current cathodic protection system	116
6.3.1 Cathodic protection anode systems	120
6.3.2 Deck anode systems	120

6.3.3 Anodes for vertical and soffit surfaces	124
6.3.4 Conductive coatings	128
6.3.5 Thermal sprayed zinc	128
6.3.6 Clamp on systems	133
6.3.7 Sacrificial anode systems	133
6.4 Cathodic protection system design	136
6.4.1 Choosing the anode	136
6.4.2 Transformer/rectifiers and control systems	139
6.4.3 Monitoring probes	141
6.4.4 Zone design	143
6.5 Control criteria	143
6.6 System installation	147
6.6.1 Patching for cathodic protection	147
6.6.2 Rebar connections and continuity	148
6.6.3 Monitoring probe installation	149
6.6.4 Anode installation	149
6.6.5 Transformer/rectifier and control system installation	150
6.6.6 Initial energizing	150
6.6.7 Commissioning	150
6.6.8 Operation and maintenance	151
6.7 Cathodic protection of prestressed concrete	151
6.8 Cathodic protection of epoxy coated reinforcing steel	152
6.9 Cathodic protection of structures with ASR	153
6.10 Chloride removal	154
6.10.1 Anode types	154
6.10.2 Electrolytes	154
6.10.3 Operating conditions	155
6.10.4 End point determination	157
6.10.5 Possible effects	159
6.10.6 Alkali–silica reactivity	159
6.10.7 Bond strength	159
6.10.8 Results after treatment: beneficial effects of passing currents through concrete	161
6.11 Realkalization	161
6.11.1 Anode types	162
6.11.2 Electrolytes	162
6.11.3 Operating conditions	162
6.11.4 End point determination	163
6.11.5 Possible effects	163
6.12 Comparison of techniques	163
6.12.1 Advantages of all electrochemical techniques	163
6.12.2 Disadvantages of all electrochemical techniques	164
6.12.3 Cathodic protection	164
6.12.4 Electrochemical chloride extraction	164

	6.12.5 Realkalization	164
	6.12.6 Costs	164
	References	165

7 Rehabilitation methodology — 169
7.1 Technical differences between repair options — 171
7.2 Repair costs — 171
7.3 Carbonation options — 174
    7.3.1 Patching and coating — 176
    7.3.2 Why choose realkalization? — 177
    7.3.3 Why choose corrosion inhibitors? — 179
7.4 Summary of options for carbonation repairs — 180
7.5 Chloride options — 181
    7.5.1 Patching and sealing — 181
    7.5.2 Why choose cathodic protection? — 182
    7.5.3 Why choose chloride extraction? — 183
    7.5.4 Other chloride repair options — 184
7.6 Summary — 185
References — 186

8 Understanding and calculating the corrosion of steel in concrete — 187
8.1 Initiation time T_0, carbonation induced corrosion — 188
    8.1.1 Parrott's determination of carbonation rates from permeability — 189
8.2 Chloride ingress rates (initiation) — 189
    8.2.1 The parabolic approximation — 190
    8.2.2 Sampling variability for chlorides — 191
    8.2.3 Mechanisms other than diffusion — 191
8.3 Rate of depassivation (activation) — 192
8.4 Deterioration and corrosion rates, T_1 — 192
    8.4.1 The Clear/Stratfull empirical calculation — 194
8.5 Corrosion without spalling — 195
8.6 Pitting corrosion — 195
8.7 Cracking and spalling rates, condition indices and end of functional service life — 195
8.8 Summary of methodology to determine service life — 197
References — 197

9 Building for durability — 199
9.1 Cover, concrete and design — 199
9.2 Fusion bonded epoxy coated rebars — 201
    9.2.1 How does epoxy coating work? — 203
    9.2.2 Problems with epoxy coating — 206

	9.2.3 Advantages and disadvantages of fusion bonded epoxy coated rebars	207
9.3	Waterproofing membranes	208
	9.3.1 Advantages and disadvantages of waterproofing membranes	210
9.4	Penetrating sealers	210
9.5	Galvanized rebar	211
9.6	Stainless steel reinforcement	211
9.7	Corrosion inhibitors	212
9.8	Installing cathodic protection in new structures	212
9.9	Durable buildings	213
9.10	Conclusions	214
	References	215

10 Future developments — 216

Appendix A Bodies involved in corrosion and repair of reinforced concrete — 220

Appendix B Strategic Highway Research Program: published reports on concrete and structures (concrete and the corrosion of steel in atmospherically exposed reinforced concrete bridge components suffering from chloride induced corrosion) — 222

Appendix C Strategic Highway Research Program: unpublished reports on concrete and structures (concrete and the corrosion of steel in atmospherically exposed reinforced concrete bridge components suffering from chloride induced corrosion) — 231

Index — 239

Preface

This book provides information on corrosion of steel in atmospherically exposed concrete structures to guide those responsible for buildings, bridges and all reinforced concrete structures as they design, construct and maintain them. It reviews the present state of knowledge of corrosion of steel in concrete from the theory, through to site investigations and remedying the problems. There is also a section on building structures to be more corrosion resistant.

The aim of the book is to educate and guide engineers, surveyors, students and owners of structures so that they will have a clearer idea of the problem of corrosion, its causes and where to go to start finding solutions. No single book can be a complete course in any subject and there is no substitute for personal 'hands on' experience. There are many experienced engineers and corrosion experts who deal with these problems on a day to day basis. The reader is recommended to seek expert advice when dealing with subjects that are new to him. However, the aim of being an 'informed client' is to be recommended and this book should help in that respect. It is also essential that students of civil engineering and building sciences understand the problems they will face as they start to practise their skills. We need more engineers, materials scientists and surveyors who are trained to look at wider durability issues than the purely structural.

While corrosion of steel in concrete is a major cause of deterioration, it is not the only one. Out in the real world we must not become blinkered to other problems like alkali–silica reactivity, freeze–thaw damage and the structural implications of the damage done and of repairs. In this book, however, we concentrate on the corrosion issue, although there will be passing references to other problems where relevant.

It is impossible in one book to be totally comprehensive so some subjects such as concrete chemistry and additives to concrete have not been covered in detail; these are better covered in other specialist texts. The book concentrates on corrosion itself and how it is dealt with by methods sometimes unique to the corrosion field.

Acknowledgements

It is difficult to know where to start and end with acknowledgements. I started my career after finishing university at the Central Electricity Research Laboratory, working in one of the largest groups of corrosion scientists in the UK and Europe. Sadly that group is dispersed with the privatization of the Central Electricity Generating Board. I learned the basics of corrosion from Mike Manning, Jonathan Forrest and Ed Metcalfe and many others at CERL.

I then joined Taywood Engineering and enjoyed carrying out some of the first trials of cathodic protection on reinforced concrete structures above ground in the UK, Hong Kong and Australia. I learned enormous amounts about deterioration of reinforced concrete structures, civil engineering and contracting from Roger Browne, Roger Blundell, Roger McAnoy, Phil Bamforth and a powerful team of engineers and scientists who were all at the forefront of concrete technology. I continue to have good relations and helpful exchanges with the Taywood team.

I was then privileged to be invited to work at the Strategic Highway Research Program (SHRP) in Washington, DC. For three years we coaxed a $150 million research programme into life, and after five years we then shut it all down having spent the allotted budget. My thanks go to Damian Kulash who led the programme and was brave enough to take on an unknown foreigner to oversee the structures research and to Jim Murphy and Don Harriott who led the Concrete and Structures Group and taught me about highways. I must also thank all my colleagues at SHRP, the researchers who did the real research work and the advisory and expert committee members who gave their time and insights to help create some very valuable manuals and guidance for highway engineers.

And so to the present. After five years as an independent consultant I have had the chance to work with some of the leading experts in the field of corrosion and deterioration of reinforced concrete. I have picked their brains mercilessly and enjoyed working and relaxing with many such as Brian Wyatt of Tarmac; Nick Buenfeld and Gareth Glass of Imperial College; Ken Boam, Mike Gower and Peter Johnson of Maunsells; David Whiting of Construction Technology Laboratories; Ken Clear and Jack Bennett; Carmen Andrade of the Instituto Eduardo Torroja in Madrid; Jesús Rodríguez of Geocisa; and many more. I must also thank all those who provided the pictures used to illustrate the text. I have tried to acknowledge the sources but I offer apologies to any I have overlooked.

My thanks also go to my publisher, and to Nick Clarke whose helpful suggestions have strengthened the book considerably, and to my reviewers Steve Millard, John Miller and Peter Pullar Strekker. They helped clarify and extend the text and the time they put in was considerable. However, this work is my own and I am responsible for any errors and inconsistencies that you the reader may find in it.

Finally, eternal thanks to my wife, Veronica, and my parents, Olive and Philip. Everything one says about the support of the family sounds glib and clichéd, but it is, none the less, true. Without my parents' support I would not have got where I am today. Veronica's support has been invaluable in quietly encouraging me to get on with the job, make sure that I am focused on the task in hand not involved in pie in the sky ideas. She has supported my activities while getting on with her own very busy career.

My apologies for naming such a small number of the many people whom I have had the pleasure and privilege of working with. My thanks to all of them and their willingness to share their knowledge with me and many others.

Glossary

These definitions are not full, accurate scientific or dictionary definitions and may be incomplete if used outside the context of the subject of corrosion of steel in concrete.

Acid A solution that (among other things) attacks steel and other metals and reacts with alkalis forming a neutral product and water.

Alkali A solution that (among other things) protects steel and other metals from corrosion and reacts with acids forming a neutral product and water.

Anode
1. The site of corrosion in an aqueous corrosion cell (a combination of anodes and cathodes).
2. An external component introduced into a cathodic protection system to be the site of the oxidation (q.v.) reaction and prevent corrosion of the metal object to be protected.
3. The positive pole of a simple electrical cell (battery).

Carbonation The process by which carbon dioxide (CO_2) in the atmosphere reacts with water in concrete pores to form carbonic acid and then reacts with the alkalis (q.v.) in the pores, neutralizing them. This can then lead to the corrosion of the reinforcing steel.

Cathode
1. The site of a charge balancing reaction in a corrosion cell.
2. The protected metal structure in a cathodic protection system.
3. The negative pole of a simple electrical cell (battery).

Cathodic protection A process of protecting a metal object or structure from corrosion by the installation of a sacrificial anode (q.v.)

or impressed current system that makes the protected object a cathode (q.v.) and thus resistant to corrosion.

Cathodic protection anode A cathodic protection anode for steel in concrete can be a conductive paint or other conductive material that will adhere to concrete, or a metal mesh or other conductive material that can be embedded in a concrete overlay on the surface of the structure to be protected. Anodes may be impressed current or sacrificial.

Cement (paste) Portland cement is a mixture of alumina, silica, lime, iron oxide and magnesia ground to a fine powder, burned in a kiln and ground again. Cement paste is the binding agent for mortar and (Portland cement) concrete after hydration.

Chloride The negative ion (q.v.) in salt (sodium chloride), found in sea salt, deicing salt and calcium chloride admixture for concrete. Chloride ions promote corrosion of steel in concrete but are not used up by the process so they can concentrate and accelerate corrosion.

Chloroaluminates Chemical compounds formed in concrete when chlorides combine with the C_3A in the hardened cement paste. These chlorides are no longer available to cause corrosion. Sulphate resisting cements have a low C_3A content and are more prone to chloride induced corrosion than normal Portland cement based concretes.

Concrete Ordinary Portland cement concrete is a mixture of cement (q.v.), fine and coarse aggregates and water. The water reacts with the cement to bind the aggregates together.

Corrosion The process by which a refined metal reverts back to its natural state by an oxidation reaction with the non-metallic environment (e.g. oxygen and water).

Galvanic action The consequence of coupling together two dissimilar metals in a solution. The most active one will become an anode and corrode. A current will pass between them. Galvanic action drives electric batteries (single cells). It is used to stop corrosion by applying zinc to steel (galvanizing), and in sacrificial anode cathodic protection.

Half cell Usually a pure metal in a solution of (fixed) concentration. The half reaction of the metal ions dissolving and reprecipitating creates a steady potential when linked to another half cell. Two half cells make an electrochemical cell that can be a model for corrosion or be used to generate electricity (a single cell, often called a battery). Reference half cells are connected to reinforcing steel to measure 'corrosion potentials' that show the corrosion condition of the steel in the concrete.

Impressed current cathodic protection A method of cathodic protection that uses a power supply and an inert (or controlled consumption) anode (q.v.) to protect a metallic object or structure by making it the cathode (q.v.).

Incipient anode An area of steel in a corroding structure that was originally cathodic due to the action of a local anode (q.v.). When the local anode is treated by patch repairing, the incipient anode is no longer protected and starts to corrode.

Ion An atom or molecule with electrons added or subtracted. Ionic compounds like salt (sodium chloride) are composed of balanced ions ($NaCl = Na^+ + Cl^-$). Some ions are soluble (e.g. Na^+, Cl^-, Fe^{2+}) which can be important for transport through concrete.

Ionic current An electric current that flows as ions through an aqueous medium (e.g. concrete pore water), as opposed to an electronic flow of electrons through a metal conductor.

iR drop Electrical current passing through a solution of finite resistance generates a voltage. This is superimposed on the half cell (q.v.) potential and must be subtracted to get accurate readings in linear polarization and in cathodic protection. This is most easily done by 'instant off' measurements of potentials taken within a few seconds of switching off the current.

Oxidation The process of removing electrons from an atom or ion. The process:

$$Fe \rightarrow Fe^{2+} + 2e^-$$
$$Fe^{2+} \rightarrow Fe^{3+} + e^-$$

is the oxidation of iron to its ferrous (Fe^{2+}) and ferric (Fe^{3+}) oxidation state. Oxidation is done by an oxidizing agent, of which oxygen is only one of many.

Passivation
The process by which steel in concrete is protected from corrosion by the formation of a passive layer due to the highly alkaline environment

created by the pore water. The passive layer is a thin, dense layer or iron oxides and hydroxides with some mineral content, that is initially formed as bare steel is exposed to oxygen and water, but then protects the steel from further corrosion as it is too dense to allow the water and oxygen to reach the steel and continue the oxidation process.

pH A measure of acidity and alkalinity based on the fact that the concentration of hydrogen ions $[H^+]$ (acidity) times hydroxyl ions $[OH^-]$ (alkalinity) is 10^{-14} moles/l in aqueous solutions:

$$[H^+][OH^-] = 1 \times 10^{-14}$$
$$pH = -\log[H^+]$$
$$pH + pOH = 14$$

i.e. a strong acid has pH = 1 (or less), a strong alkali has pH = 14 (or more), a neutral solution has pH = 7. Concrete has a pH of 12 to 13. Steel corrodes at pH 10 to 11.

Prestressing The process of applying compressive stress to concrete using steel rods, wires or bars to make stronger elements than by conventional reinforcement. There are two forms of prestressing; pretensioned and post-tensioned. Pretensioned structures and elements are made by applying a load to the steel, casting the concrete around it and then, when it has hardened and developed sufficient strength, releasing the load on the steel which is then taken up by the concrete.

Post-tensioning is the process of casting the concrete with ducts in. After the concrete has hardened and developed sufficient strength cables in the ducts are tensioned and anchored. Ducts may be filled with cement grout (bonded) or with protective grease (unbonded). The grouting process is vital to the corrosion protection of the post-tensioning cables.

Reduction Chemically this is the reverse of oxidation (q.v.). The incorporation of electrons into a nonmetal oxidizing agent when a metal is oxidized. When oxygen (O_2) oxidizes iron (Fe) to Fe^{2+} it receives the electrons that the iron gives up and is itself reduced:

$$O_2 + 4e^- \rightarrow 2O^{2-}$$
$$2e^- + H_2O + \tfrac{1}{2}O_2 \rightarrow 2OH^-$$

are reduction reactions.

Pore (water) Concrete contains microscopic pores. These contain alkaline oxides and hydroxides of sodium, potassium and calcium.

Water will move in and out of the concrete saturating, part filling and drying out the pores according to the external environments. The alkaline pore water sustains the passive layer if not attacked by carbonation or chlorides (q.v.).

Reference electrode An alternative name for a half cell (q.v.).

Reinforced concrete Concrete containing a network of reinforcing steel bars to make a composite material that is strong in tension as well as in compression. Smaller volumes of material can therefore be used to make beams, bridge spans, etc. compared with unreinforced concrete, brick or masonry.

Rust The corrosion product of iron and steel in normal atmospheric conditions. Chemically it is hydrated ferric oxide $Fe_2O_3.H_2O$. It has a volume several times that of the iron that was consumed to produce it.

Sacrificial anode cathodic protection A system of cathodic protection that uses a more easily corroded metal such as zinc, aluminium or magnesium to protect a steel object from corrosion. No power supply is required, but the anode (q.v.) is consumed.

SHRP The Strategic Highway Research Program. A $150 million research effort that spent about $10 million on corrosion of reinforced concrete bridges suffering from chloride induced corrosion. SHRP produced about 40 reports covering assessment, repair and rehabilitation methodology.

Steel An alloy of iron with up to 1.7% carbon to enhance its physical properties.

Titanium mesh anode A type of impressed current anode consisting of an expanded titanium mesh coated by a corrosion resistant film of mixed metal oxides. After being fixed to the concrete surface the mesh is covered with concrete or mortar.

1
Introduction

Reinforced concrete is a versatile, economical and successful construction material. It can be moulded to a variety of shapes and finishes. Usually it is durable and strong, performing well throughout its service life. However, sometimes it does not perform adequately as a result of poor design, poor construction, inadequate materials selection, a more severe environment than anticipated or a combination of these factors.

The corrosion of reinforcing steel in concrete is a major problem facing civil engineers and surveyors today as they maintain an ageing infrastructure. Potentially corrosion rehabilitation is a very large market for those who develop the expertise to deal with the problem. It is also a major headache for those who are responsible for dealing with structures suffering from corrosion.

One American estimate is that $150 billion worth of corrosion damage on their interstate highway bridges is due to deicing and sea salt induced corrosion. In a recent Transportation Research Board Report on the costs of deicing (Transportation Research Board, 1991), the annual cost of bridge deck repairs was estimated to be $50 to $200 million, with substructures and other components requiring $100 million a year and a further $50 to $150 million a year on multistorey car parks.

In the United Kingdom, the Highways Agency's estimate of salt induced corrosion damage is a total of £616.5 million on motorway and trunk road bridges in England and Wales alone (Wallbank, 1989). These bridges represent about 10% of the total bridge inventory in the country. The eventual cost may therefore be ten times the Highways Agency's estimate. The statistics for Europe, the Asian Pacific countries and Australia are similar.

In the Middle East the severe conditions of a warm marine climate with saline ground waters increase all corrosion problems. This is made worse by the difficulty of curing concrete and has led to very short lifetimes for reinforced concrete structures (Rasheeduzzafar *et al.*, 1992).

In many countries with rapidly developing infrastructures, economies in construction have led to poor quality concrete and low concrete cover to the steel resulting in carbonation problems.

Corrosion became a fact of life as soon as man started digging ores out of the ground, smelting and refining them to produce the metals that we use so widely in the manufacturing and construction industries. When man has finished refining the steel and other metals that we use, nature sets about reversing the process. The refined metal will react with the non-metallic environment to form oxides, sulphates, sulphides, chlorides, etc. which no longer have the required chemical and physical properties of the consumed metal. Billions of dollars are spent every year in protecting, repairing and replacing corrosion damage. Occasionally lives are lost when steel pipes, pressure vessels or structural elements on bridges fail. But corrosion is a slow process and is usually easy to detect before catastrophic failure, and the consequences of corrosion are usually economic rather than death or injury. The corrosion of steel in concrete has previously caused only economic problems; however, recently there have been cases of large pieces of concrete falling from bridges in North America, with the loss of a motorist's life in New York City. The failure of post-tensioning tendons in a bridge caused a collapse with the loss of a tanker driver's life in Europe.

The economic loss and damage caused by the corrosion of steel in concrete makes it arguably the largest single infrastructure problem facing industrialized countries. Our bridges, public utilities, chemical plants and buildings are ageing. Some can be replaced, others would cause great cost and inconvenience if they were taken out of commission. With major political arguments about how many more bridges, power plants and other structures we can build, it becomes crucial that the existing structures perform to their design lives and limits and are maintained effectively.

One of the biggest causes of corrosion of steel in concrete is the use of deicing salts on our highways. In the USA approximately 10 million tonnes of salt are applied per year to highways. In the UK 1 to 2 million tonnes per year are applied (but to a proportionately far smaller road network). In continental Europe the application rates are comparable, except where it is too cold for salt to be effective, or the population density too low to make salting worthwhile (in northern Scandinavia, for example). However, research in the USA has shown that the use of deicing salt is still more economic than alternative, more expensive, less effective deicers (Transportation Research Board, 1991).

In the early years of this century corrosion of steel in concrete was attributed to stray current corrosion from electric powered vehicles. It was only in the 1950s in the USA that it was finally accepted that there were corrosion problems on bridges far from power lines but in areas where sea salt or deicing salts were prevalent. During the 1960s attempts were made to quantify the problem and in the 1970s the first cathodic protection systems were installed to deal with the problem. In Europe and the Middle East sea water for mixing concrete and the addition of calcium chloride set accelerators were acceptable until the 1960s and still used until the 1970s in a mistaken belief that most of the chlorides would be bound up in the cement and would not cause corrosion. This was found to be an expensive error over the next 20 years, particularly in the Middle East, where the low availability of potable water led to the frequent use of sea water for concrete mixing.

In the UK the Concrete in the Oceans research programme in the 1970s did much to improve our understanding of the fundamentals of corrosion of steel in concrete, particularly in marine conditions (Wilkins and Lawrence, 1980). Government laboratories around the world have worked on the problems since the 1960s when the problems started to manifest themselves.

The book frequently refers to the results of the Strategic Highway Research Program (SHRP). The author was a member of the SHRP staff for the first three years and a consultant for the last three years of SHRP. About $10 million worth of research was carried out over six years (1987–93) on corrosion problems on existing reinforced concrete bridges. A complete list of SHRP reports on corrosion of steel in concrete is included in the appendix. SHRP was probably the largest single effort to address the problem of corrosion of steel in concrete and provide practical solutions for engineers to use.

The text draws on corrosion problems and experience from around the world where possible, giving a balanced view of different approaches in different countries, particularly comparing the European and North American approaches. Inevitably it concentrates on the author's primary experience in the UK, North America and, to a lesser extent Europe, the Middle East and the Pacific Rim countries.

The next two chapters following this introduction explore the basics of corrosion of steel in concrete. The author has attempted to be thorough but also to start from the position of the generalist, with a minimal memory of chemistry. A glossary of terms is provided to remind the reader of the basic terms used.

The first requirement when addressing a deterioration problem is to quantify it. Chapter 4 therefore discusses condition evaluation and the testing procedures and techniques we can use to assess the causes and extent of the corrosion damage on a structure.

Chapters 5 and 6 are then concerned with repair and rehabilitation options, first the conventional physical intervention of concrete repair, patching, overlaying and coatings. The electrochemical techniques of cathodic protection, chloride removal and realkalization are dealt with in Chapter 6.

Chapter 7 discusses how to select our rehabilitation method. It summarizes the merits and limitations of the different repair techniques covered in the previous chapters and the cost differentials. The following chapter reviews models for determining deterioration rates and service lives.

The penultimate chapter turns away from the issues of dealing with the problems on existing structures and reviews the options for new construction to ensure that the next generations of structures do not show the problems associated with some of the present generation. Many materials and methods of improving durability are on the market; which are cost effective?

The final chapter discusses future developments. As materials science, computer technology, electronics and other disciplines advance, new methods for assessing corrosion are becoming available. The author speculates about what will be on offer and perhaps more importantly what we need to efficiently assess and repair corrosion damage on our fixed infrastructure in the future.

REFERENCES

Rasheeduzzafar, Dakhil, F.H., Bader, M.A. and Khan, M.N. (1992) 'Performance of corrosion resisting steel in chloride-bearing concrete', *ACI Materials Journal*, **89** (5), 439–48.

Transportation Research Board (1991) *Highway Deicing: Comparing Salt and Calcium Magnesium Acetate*, Special Report 235, National Research Council, Washington, DC.

Wallbank, E.J. (1989) *The Performance of Concrete in Bridges: A Survey of 200 Highway Bridges*, HMSO, London.

Wilkins, N.J.M. and Lawrence, P.F. (1980) *Concrete in the Oceans: Fundamental Mechanisms of Corrosion of Steel Reinforcements in Concrete Immersed in Sea Water*, Technical Report 6, CIRIA/UEG Cement and Concrete Association, Slough, UK.

2
Corrosion of steel in concrete

This chapter discusses the basics of corrosion and how they apply to steel in concrete. The glossary of terms at the beginning of the book gives some definitions of the terms used here as they apply to the corrosion of steel in concrete.

Why does steel corrode in concrete? A more sensible question is why steel does not corrode in concrete. We know from experience that mild steel and high strength reinforcing steel bars corrode (rust) when air and water are present. As concrete is porous and contains moisture why does steel in concrete not usually corrode?

The answer is that concrete is alkaline. Alkalinity is the opposite of acidity. Metals corrode in acids, whereas they are often protected from corrosion by alkalis.

When we say that concrete is alkaline we mean that it contains microscopic pores with high concentrations of soluble calcium, sodium and potassium oxides. These oxides form hydroxides, which are very alkaline, when water is added. This creates a very alkaline condition (pH 12–13 – see the glossary for a definition of pH). The composition of the pore water and the movement of ions and gases through the pores is very important when analysing the susceptibility of reinforced concrete structures to corrosion. This is more fully discussed in Chapter 3.

The alkaline condition leads to a 'passive' layer forming on the steel surface. A passive layer is a dense, impenetrable film which, if fully established and maintained, prevents further corrosion of the steel. The layer formed on steel in concrete is probably part metal oxide/hydroxide and part mineral from the cement. A true passive layer is a very dense, thin layer of oxide that leads to a very slow rate of oxidation (corrosion). There is some discussion whether or not the layer on the steel is a true passive layer as it seems to be thick compared with other passive layers and it consists of more than just metal oxides; but it behaves like a passive layer and it is therefore generally referred to as such.

Corrosion engineers spend much of their time trying to find ways of stopping corrosion of steel by applying protective coatings. Metals such as zinc or polymers such as acrylics or epoxies are used to stop corrosive conditions getting to steel surfaces. The passive layer is the corrosion engineer's dream coating as it forms itself and will maintain and repair itself as long as the passivating (alkaline) environment is there to regenerate it if it is damaged. If the passivating environment can be maintained, it is far better than any artificial coatings such as galvanizing or fusion bonded epoxy that can be consumed or damaged, allowing corrosion to proceed in damaged areas.

However, the passivating environment is not always maintained. Two processes can break down the passivating environment in concrete. One is carbonation and the other is chloride attack. These will be discussed in Chapter 3. In the rest of this chapter we will discuss what happens when depassivation has occurred.

2.1 THE CORROSION PROCESS

Once the passive layer breaks down then areas of rust will start appearing on the steel surface. The chemical reactions are the same whether corrosion occurs by chloride attack or carbonation. When steel in concrete corrodes it dissolves in the pore water and gives up electrons:

$$\text{The anodic reaction: } Fe \rightarrow Fe^{2+} + 2e^- \qquad (2.1)$$

The two electrons ($2e^-$) created in the anodic reaction must be consumed elsewhere on the steel surface to preserve electrical neutrality. In other words, it is not possible for large amounts of electrical charge to build up at one place on the steel; another chemical reaction must consume the electrons. This is a reaction that consumes water and oxygen:

$$\text{The cathodic reaction: } 2e^- + H_2O + \tfrac{1}{2}O_2 \rightarrow 2OH^- \qquad (2.2)$$

This is illustrated in Figure 2.1. You will notice that hydroxyl ions ($2OH^-$) are generated in the cathodic reaction. These ions increase the local alkalinity and will therefore strengthen the passive layer, warding off the effects of carbonation and chloride ions at the cathode. Note that water and oxygen are needed at the cathode for corrosion to occur.

The anodic and cathodic reactions (2.1 and 2.2) are only the first steps in the process of creating rust. However, this pair of reactions is critical to the understanding of corrosion and is widely quoted in any discussion on corrosion and corrosion prevention for steel in concrete. The reactions will be referred to often in this book.

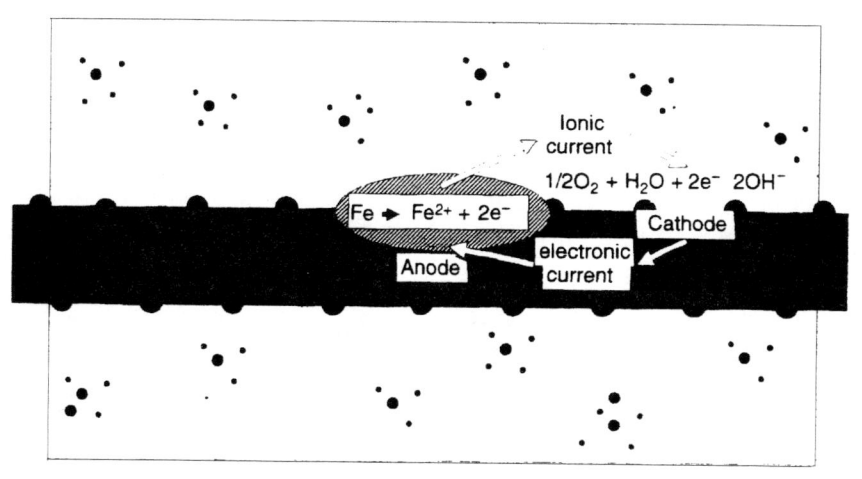

Figure 2.1 The anodic and cathodic reactions.

If the iron were just to dissolve in the pore water (the ferrous ion Fe^{2+} in equation 2.1 is soluble) we would not see cracking and spalling of the concrete. Several more stages must occur for 'rust' to form. This can be expressed in several ways; one is shown below where ferrous hydroxide becomes ferric hydroxide and then hydrated ferric oxide or rust:

$$Fe^{2+} + 2OH^- \rightarrow Fe(OH)_2 \quad (2.3)$$
$$\text{Ferrous hydroxide}$$

$$4Fe(OH)_2 + O_2 + 2H_2O \rightarrow 4Fe(OH)_3 \quad (2.4)$$
$$\text{Ferric hydroxide}$$

$$2Fe(OH)_3 \rightarrow Fe_2O_3.H_2O + 2H_2O \quad (2.5)$$
$$\text{Hydrated ferric oxide (rust)}$$

The full corrosion process is illustrated in Figure 2.2. Unhydrated ferric oxide Fe_2O_3 has a volume of about twice that of the steel it replaces when fully dense. When it becomes hydrated it swells even more and becomes porous. This means that the volume increase at the steel/concrete interface is two to ten times. This leads to the cracking and spalling that we observe as the usual consequence of corrosion of steel in concrete and the red/brown brittle, flaky rust on the bar and the rust stains seen at cracks in the concrete. The cracking and spalling process is discussed more fully in Section 3.3 on corrosion damage.

Several factors in the explanation given in this section are important and will be used later to explain how we measure and stop corrosion.

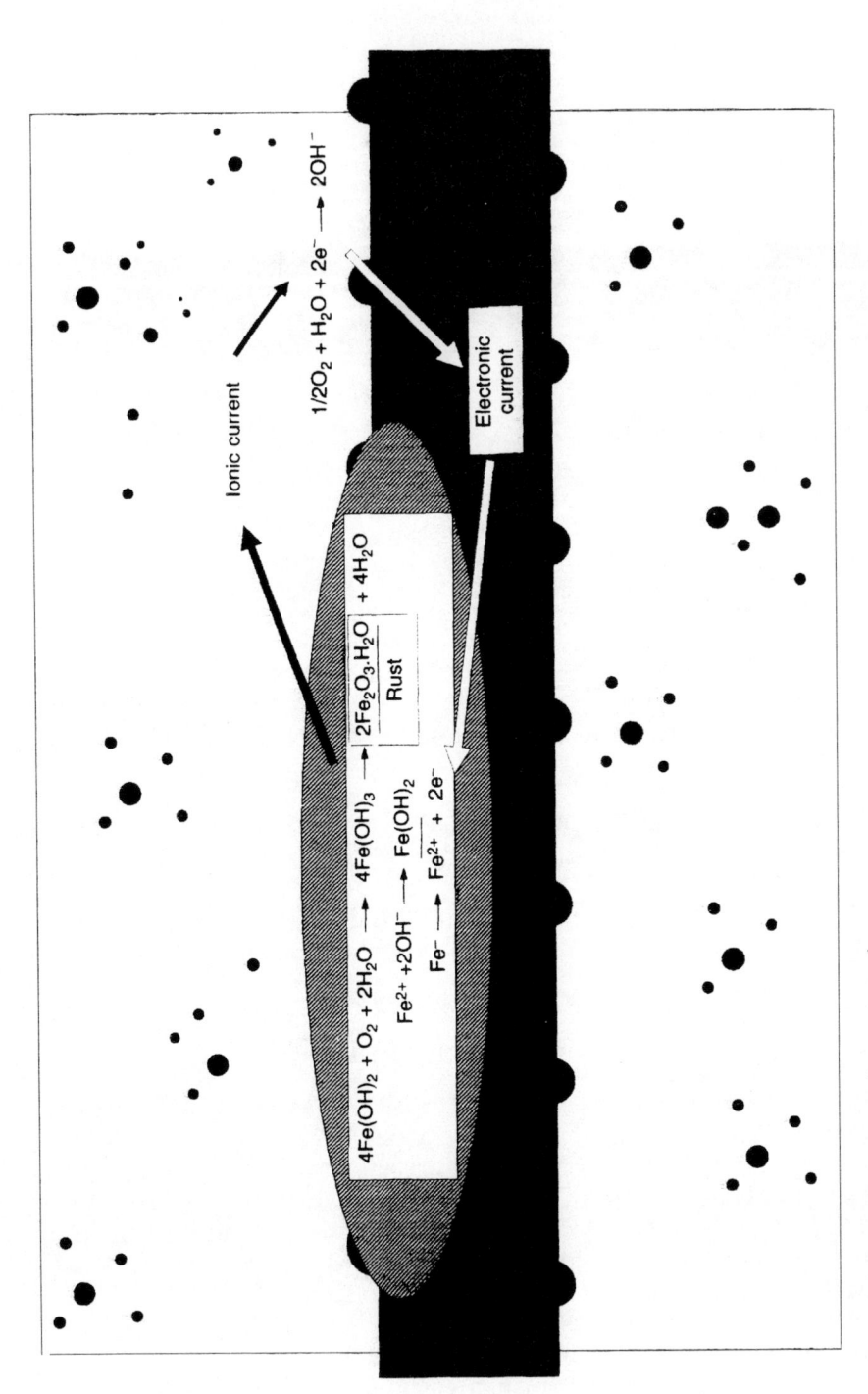

Figure 2.2 The corrosion reactions on steel.

The electrical current flow, and the generation and consumption of electrons in the anode and cathode reactions are used in half cell potential measurements and cathodic protection. The formation of protective, alkaline hydroxyl ions is used in cathodic protection, electrochemical chloride removal and realkalization. The fact that the cathodic and anodic reactions must balance each other for corrosion to proceed is used in epoxy coating protection of rebars. This will be discussed later.

2.2 BLACK RUST

There is an alternative to the formation of 'normal' red rust described in reactions 2.3 to 2.5 above. If the anode and cathode are well separated (by several hundred millimetres) and the anode is starved of oxygen (say by being underwater) the iron as Fe^{2+} will stay in solution. This means that there will be no expansive forces to crack the concrete as described above and in Section 3.3 so corrosion may not be detected.

This type of corrosion (known as 'black' or 'green' rust due to the colour of the liquid seen on the rebar when first exposed to air after breakout) is found under damaged waterproof membranes and in some underwater or other water saturated conditions. It is potentially

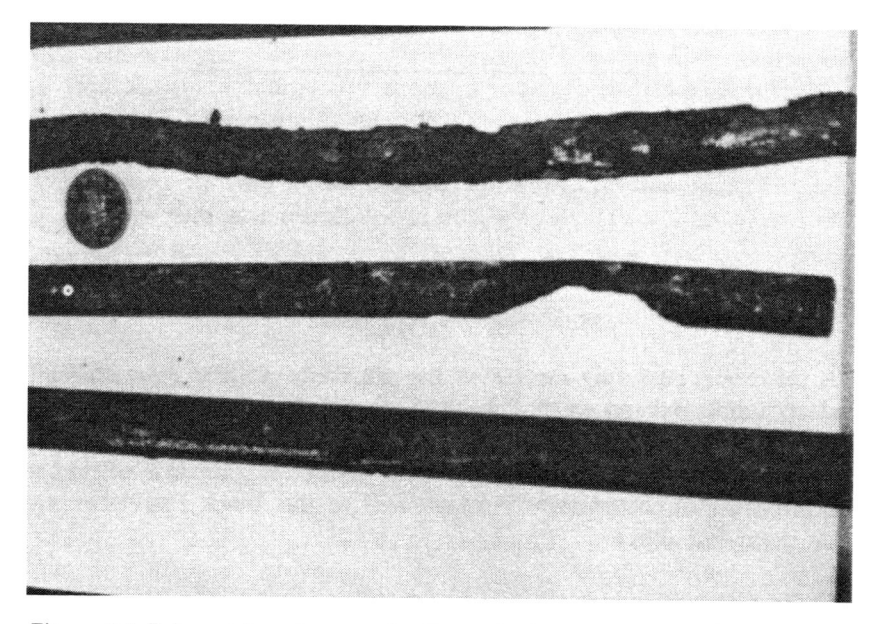

Figure 2.3 Rebars taken from under the end of a waterproof membrane. They have been subjected to low oxygen conditions and therefore show local 'wasting' of bars with no expansive oxide growth.

dangerous as there is no indication of corrosion by cracking and spalling of the concrete and the reinforcing steel may be severely weakened before corrosion is detected. Rebars may be hollowed out in such deoxygenated conditions, particularly under membranes or when water is permanently ponded on the surface.

Examples of rebars attacked in this way are shown in Figure 2.3. These bars were taken from underneath damaged waterproof membranes. Rust staining on the concrete surface may be indicative of this type of attack, but obviously if water is getting under a membrane and excluding oxygen it is unlikely that the iron in solution will get to the concrete surface where it will then precipitate out to form rust stains.

2.3 PITS, STRAY CURRENT AND BACTERIAL CORROSION

2.3.1 Pit formation

Corrosion of steel in concrete generally starts with the formation of pits. These increase in number, expand and join up leading to the generalized corrosion usually seen on reinforcing bars exposed to carbonation or chlorides. The formation of a pit is illustrated in Figure 2.4.

The chemistry of pitting is quite complex and is explained in most corrosion textbooks. However, the principle is fairly simple, especially where chlorides are present. At some suitable site on the steel surface (often thought to be a void in the cement paste or a sulphide inclusion in the steel), the passive layer is more vulnerable to attack and an electrochemical potential difference attracts chloride ions. Corrosion is initiated and acids are formed; hydrogen sulphide from the sulphide (MnS) inclusion and HCl from the chloride ions if they are present. Iron dissolves (equation 2.1), and the iron in solution reacts with water:

$$Fe^{2+} + H_2O \rightarrow FeOH^+ + H^+ \qquad (2.6)$$

$$MnS + 2H^+ \rightarrow H_2S + Mn^{2+} \qquad (2.7)$$

A pit forms, rust may form over the pit, concentrating the acid (H^+), and excluding oxygen so that the iron stays in solution preventing the formation of a protective oxide layer and accelerating corrosion.

We will return to the subject of pitting corrosion later. It is related to the problems of coated reinforcement and to the 'black rust' phenomenon discussed above.

2.3.2 Bacterial corrosion

Another complication comes from bacterial corrosion. Bacteria in the soil (thiobacilli) convert sulphur and sulphides to sulphuric acid. This

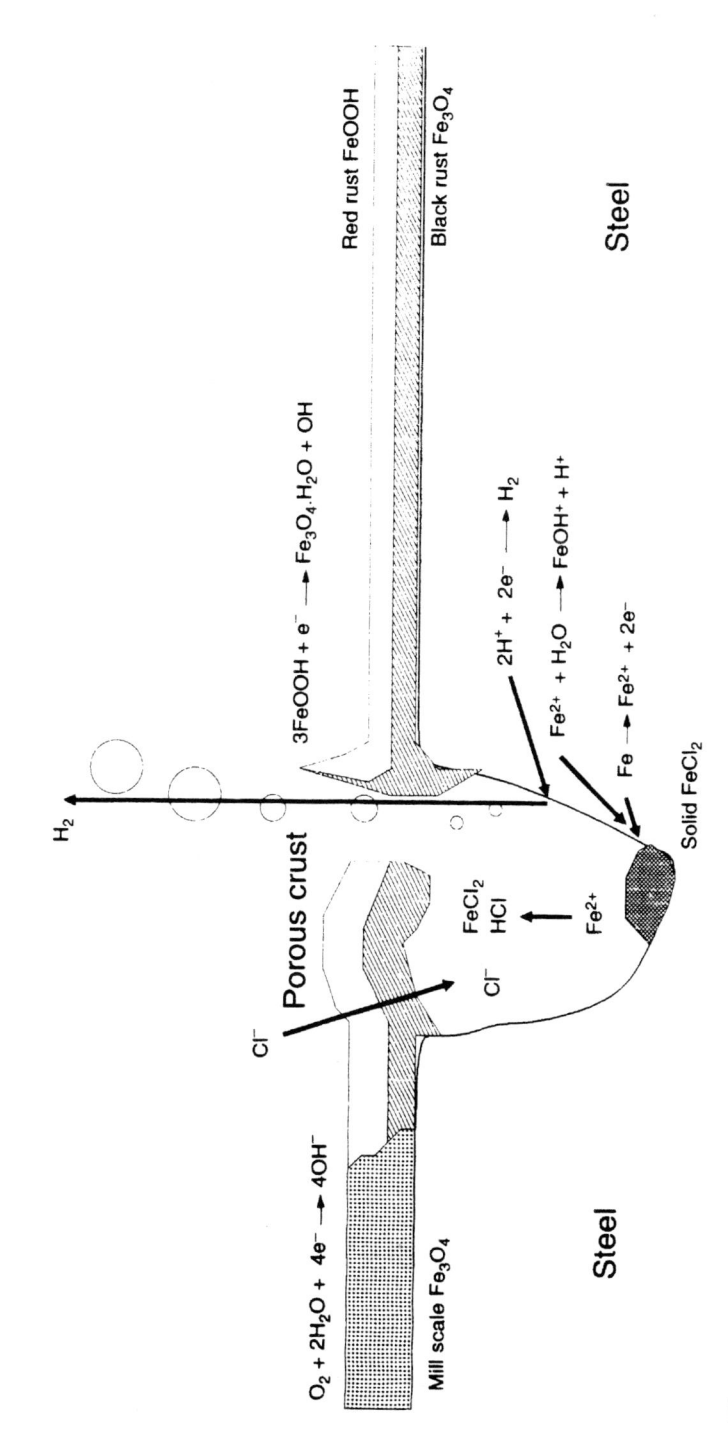

Figure 2.4 The classical corrosion model of pitting attack.

acid will attack steel causing corrosion. There are other species (ferrobacilli) that attack the sulphides in steel (FeS). This is often associated with a smell of hydrogen sulphide (rotten eggs) and smooth pitting, with a black corrosion product when rebars are exposed having been in water saturated conditions. In anaerobic (oxygen starved) conditions such bacteria can contribute to the pitting corrosion discussed above.

2.3.3 Stray current induced corrosion

As stated in the introduction, stray currents were originally blamed for all corrosion in concrete in the USA, until the problem of chloride attack was identified in the 1950s. The main cause of stray current induced corrosion was the direct current flowing through reinforcing steel due to DC traction systems on trams (streetcars) and electric trains. When current has to jump from one metal conductor to another through an ionic medium (e.g. from one reinforcement cage to another via the concrete pore water), then one end becomes negative (a cathode, i.e. cathodically protected), while the other end becomes an anode and corrodes. The problem of stray current corrosion within a cathodic protection system and an illustration of the mechanism is given in Chapter 6 where the problems of current flow through isolated rebars is discussed for electrochemical techniques.

Today stray current corrosion in bridges, buildings and above-ground structures is a specialized problem dealt with by engineers designing and maintaining light railway systems. Below ground it is the province of the cathodic protection engineer, where interaction between a cathodic protection system on pipelines, etc. can induce corrosion on adjacent buried structures. Stray current induced corrosion does not occur on AC systems, and is most unlikely on above-ground structures when the source is below ground.

2.3.4 Local versus general corrosion (macrocells versus microcells)

Corrosion is often local, with a few centimetres of corrosion and then up to a metre of clean passive bar, particularly for chloride induced corrosion. This indicates the separation of the anodic reaction (2.1) and the cathodic reaction (2.2), to form a 'macrocell'. Chloride induced corrosion gives rise to particularly well defined macrocells. This is partly due to the mechanism of chloride attack, with pit formation and with small concentrated anodes being 'fed' by large cathodes. It is also because chloride attack is usually associated with high levels of moisture giving low electrical resistance in the concrete and easy transport of ions so the anodes and cathodes can separate easily.

In North America, this is used as a way of measuring corrosion by measuring 'macrocell currents'. This is discussed later in Section 4.11.

2.4 ELECTROCHEMISTRY, CELLS AND HALF CELLS

The first two reactions we discussed in this chapter were the anodic and cathodic reactions for steel in concrete. The terms 'anode' and 'cathode' come from electrochemistry which is the study of the chemistry of electrical cells. Figure 2.5 is a basic Daniell cell which is used at high school to illustrate how chemical reactions produce electricity. The cell is composed of two 'half cells', copper in copper sulphate and zinc in zinc sulphate. The total voltage of the cell is determined by the metals used and by the nature and composition of the solutions. What is happening is that in each half cell the metal is dissolving and ions are precipitating, i.e.

$$M \longleftrightarrow M^{n+} + ne^-$$

Copper is more resistant to this reaction than zinc so when we connect the two solutions by a semi-permeable membrane (which

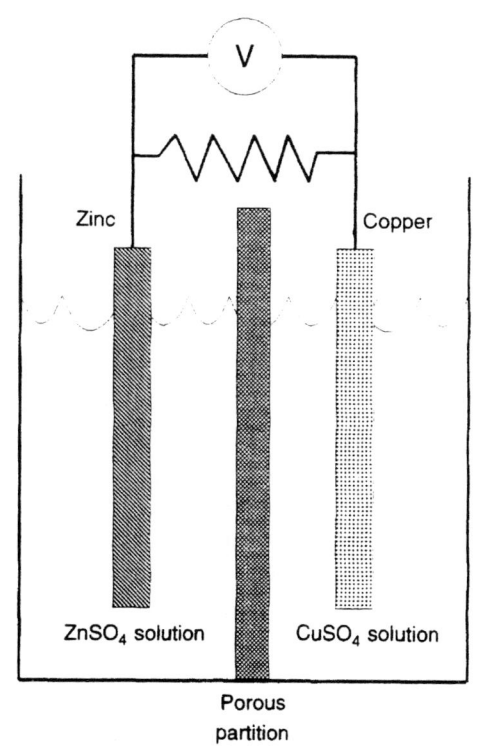

Figure 2.5 The Daniell cell. Cell voltage = 0.34 − (−0.76) = 1.10 V.

Table 2.1 Half cell potentials

$Zn \rightarrow Zn^{2+} + 2e^-$	-0.76 V
$Fe \rightarrow Fe^{2+} + 2e^-$	-0.44 V
$Cu \rightarrow Cu^{2+} + 2e^-$	$+0.34$ V

allows charge to be exchanged through it but the ions cannot pass through) and connect the two metals with a wire, the zinc goes into solution, and the copper from the copper sulphate solution plates out (is deposited) on the copper electrode.

The voltage of any half cell can be recorded against a standard hydrogen electrode (half cell). Table 2.1 gives the standard half cell potentials that are of interest to us as we evaluate corrosion problems. Half cell potentials are a function of concentration as well as the metal and the solution. A more concentrated solution is (generally) more corrosive than a dilute one so a current will flow in a cell made up of a single metal in two different concentrations of the same solution. We can consider the corrosion of steel in concrete as a concentration cell.

We can measure the corrosion risk in that cell by introducing an external half cell. This is most easily illustrated by a copper/copper sulphate half cell (see Section 4.7) moved along the surface of the

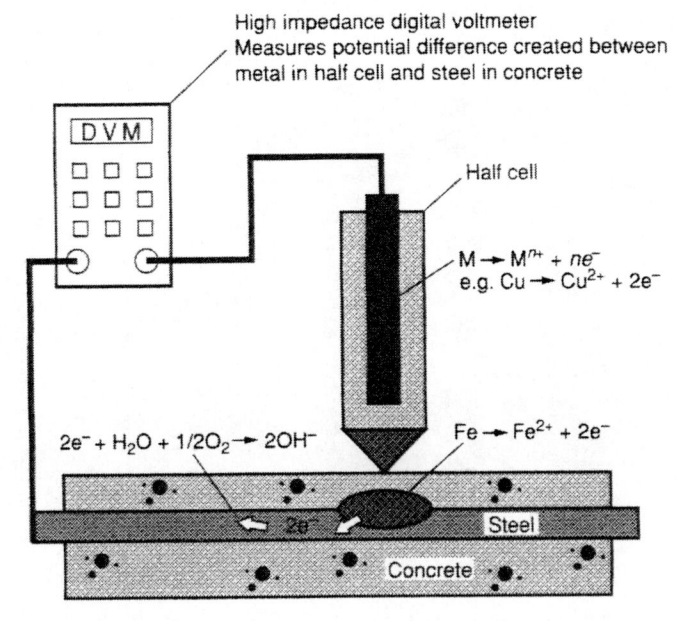

Figure 2.6 Half cell measurement of corrosion potential.

concrete which contains a rebar with anodic (corroding) areas and cathodic (passive) areas on it.

As we see in Figure 2.6, by placing a half cell on the concrete surface and connecting it via a voltmeter to the steel, we have a similar circuit to our Daniell cell (Figure 2.5). The electrical potential difference will be a function of the iron in its pore water environment. If we move the cell along the steel we will see different potentials because the iron is in different environments. At the anode it can easily go into solution like the zinc in our Daniell cell. At the cathode, the passive layer is still strong and being strengthened further by the cathodic reaction, so the steel resists dissolution.

As a result we see higher potentials (voltages) on our voltmeter in anodic, corroding areas and lower voltages in the cathodic, passive areas. The use and interpretation of half cell measurements is discussed further in Section 4.7.

We must be careful in our use of electrochemical theory to explain what is going on in a corrosion cell. Electrochemical theory generally applies to equilibrium conditions and well defined solutions. Corrosion is not an equilibrium, but a dynamic situation. The theory and equations of electrochemistry are therefore an approximation and can lead to errors if the model is stretched too far.

2.5 CONCLUSIONS

This chapter has discussed in detail the mechanism of what happens at the steel surface. The chemical reactions, formation of oxides, pitting, stray currents, bacterial corrosion, anodes, cathodes and half cells have been reviewed. The next chapter will discuss the processes that lead to the corrosion and the consequences of it in terms of damage to structures, before moving on to the measurement of the problem and how to deal with it.

3
Causes and mechanisms of corrosion and corrosion damage in concrete

There are two main causes of corrosion of steel in concrete. This chapter will discuss how chloride attack and carbonation lead to corrosion and how the corrosion proceeds once it has started. The mechanism of corrosion damage is explained. There will also be discussion of the variations that can be found when carrying out investigations in the field.

The main causes of corrosion of steel in concrete are chloride attack and carbonation. These two mechanisms are unusual in that they do not attack the integrity of the concrete. Instead, aggressive chemical species pass through the pores in the concrete and attack the steel. This is unlike normal deterioration processes due to chemical attack on concrete. Other acids and aggressive ions such as sulphate destroy the integrity of the concrete before the steel is affected. Most forms of chemical attack are therefore concrete problems before they are corrosion problems. Carbon dioxide and the chloride ion are very unusual in penetrating the concrete without significantly damaging it. Accounts of (for instance) acid rain causing corrosion of steel embedded in concrete are unsubstantiated. Only carbon dioxide and the chloride ion have been shown to attack the steel and not the concrete.

3.1 CARBONATION

Carbonation is the result of the interaction of carbon dioxide gas in the atmosphere with the alkaline hydroxides in the concrete. Like many other gases carbon dioxide dissolves in water to form an acid. Unlike most other acids the carbonic acid does not attack the cement paste, but just neutralizes the alkalis in the pore water, mainly forming calcium carbonate that lines the pores:

$$CO_2 + H_2O \rightarrow H_2CO_3 \quad (3.1)$$
Gas Water Carbonic acid

$$H_2CO_3 + Ca(OH)_2 \rightarrow CaCO_3 + 2H_2O \quad (3.2)$$
Carbonic Pore
acid solution

There is a lot more calcium hydroxide in the concrete pores than can be dissolved in the pore water. This helps maintain the pH at its usual level of around 12 or 13 as the carbonation reaction occurs. However, eventually all the locally available calcium hydroxide reacts, precipitating the calcium carbonate and allowing the pH to fall to a level where steel will corrode. This is illustrated in Figure 3.1. There is a discussion of the pH levels involved and their measurement in Section 4.8.

Carbonation damage occurs most rapidly when there is little concrete cover over the reinforcing steel. Carbonation can occur even when the concrete cover depth to the reinforcing steel is high. This may be due to a very open pore structure where pores are well connected together and allow rapid CO_2 ingress. It may also happen when alkaline reserves in the pores are low. These problems occur when there is a low cement content, high water cement ratio and poor curing of the concrete.

A carbonation front proceeds into the concrete roughly following the laws of diffusion. These are most easily defined by the statement that the rate is inversely proportional to the thickness:

$$dx/dt = D_0/x \quad (3.3)$$

where x is distance, t is time and D_0 is the diffusion constant.

The diffusion constant D_0 is determined by the concrete quality. At the carbonation front there is a sharp drop in alkalinity from pH 11–13 down to less than pH 8 (Figure 3.1). At that level the passive layer, which we saw in the previous chapter was created by the alkalinity, is no longer sustained so corrosion proceeds by the general corrosion mechanism as described.

Many factors influence the ability of reinforced concrete to resist carbonation induced corrosion. The carbonation rate, or rather the time to carbonation induced corrosion, is a function of cover thickness, so good cover is essential to resist carbonation. As the process is one of neutralizing the alkalinity of the concrete, good reserves of alkali are needed, i.e. a high cement content. The diffusion process is made easier if the concrete has an open pore structure. On the macroscopic scale

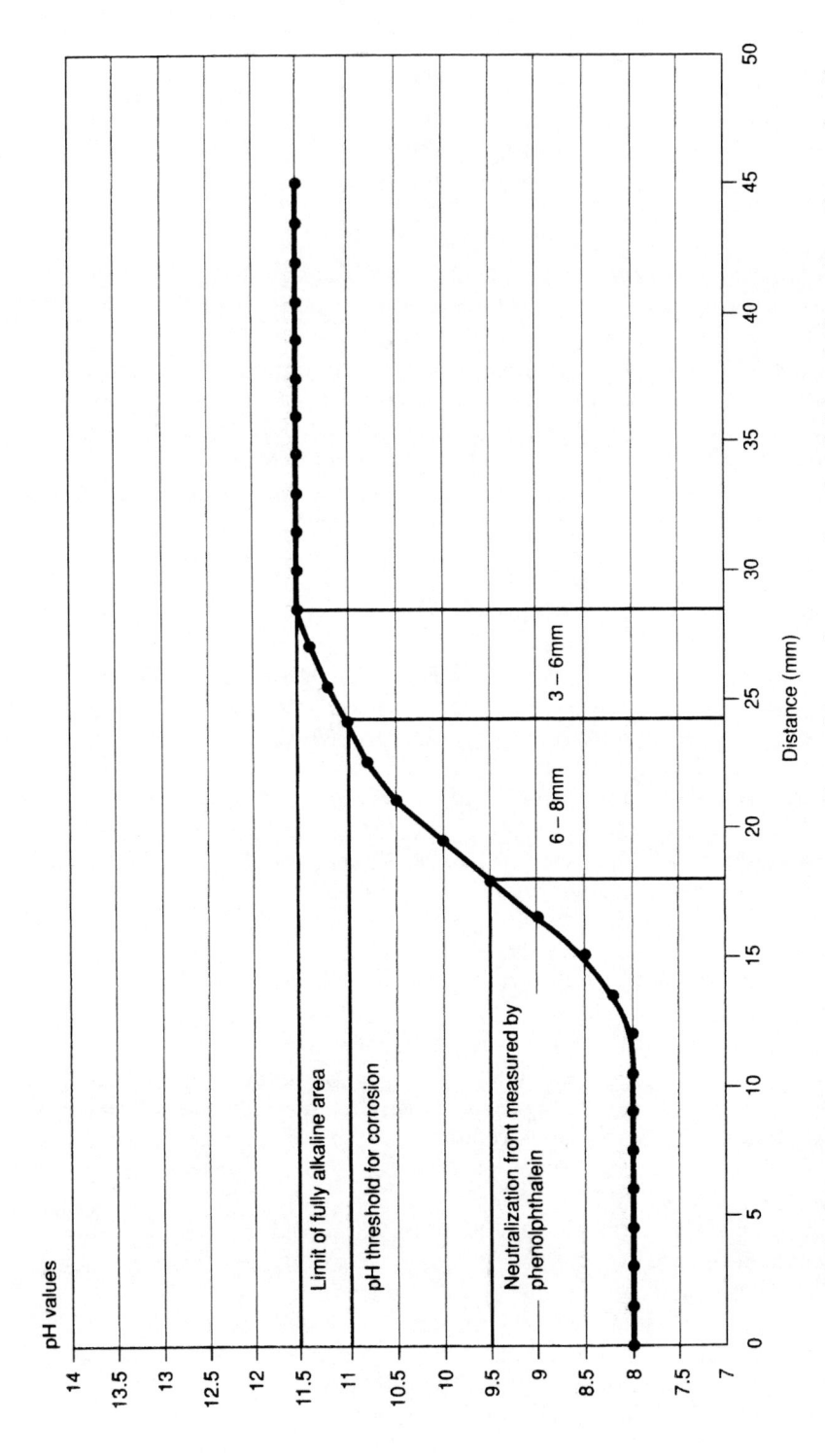

Figure 3.1 Schematic of the carbonation front and its relationship to the corrosion threshold and the colour change for phenolphthalein.

this means that there should be good compaction. On a microscopic scale well cured concrete has small pores and lower connectivity of pores so the CO_2 has a harder job moving through the concrete. Microsilica and other additives can block pores or reduce pores sizes.

Carbonation is common in old structures, badly built structures (particularly buildings) and reconstituted stone elements containing reinforcement that often have a low cement content and are very porous. Carbonation is rare on modern highway bridges and other civil engineering structures where water/cement ratios are low, cement contents are high with good compaction and curing, and there is enough cover to prevent the carbonation front advancing into the concrete to the depth of the steel within the lifetime of the structure. On those structures exposed to sea water or deicing salts, the chlorides usually penetrate to the reinforcement and cause corrosion long before carbonation becomes a problem. Wet/dry cycling on the concrete surface will accelerate carbonation by allowing carbon dioxide gas in during the dry cycle and then supplying the water to dissolve it in the wet cycle (equation 3.1). This gives problems in some countries in tropical or semi-tropical regions where the cycling between wet and dry seasons seems to favour carbonation, e.g. Hong Kong and some Pacific Rim countries.

When a repairer talks of repairing corrosion due to 'low cover' he usually means that the concrete has carbonated around the steel leading to corrosion. As the cover is low it was a quick process, perhaps within five years of construction. If the concrete were of the highest quality carbonation may not have been possible and low cover might not have mattered.

Carbonation is easy to detect and measure. A pH indicator, usually phenolphthalein in a solution of water and alcohol, will detect the change in pH across a freshly exposed concrete face. Phenolphthalein changes from colourless at low pH (carbonated zone) to pink at high pH (uncarbonated concrete). Measurements can be taken on concrete cores, fragments and down drilled holes. Care must be taken to prevent dust or water from contaminating the surface to be measured but the test, with the indicator sprayed on to the surface, is cheap and simple. Figure 3.1 shows a typical carbonation front in concrete with the pH of the concrete before and after carbonation, the pH threshold of corrosion for steel and the pH change of phenolphthalein taken from Parrott (1987). The measurement of the carbonation front is covered in Section 4.6.

3.1.1 Carbonation transport through concrete

Carbon dioxide diffuses through the concrete and the rate of movement of the carbonation front approximates to Fick's law of diffusion. This states that the rate of movement is inversely proportional to the distance from the surface as in equation 3.3 above. However, as the carbonation process modifies the concrete pore structure as it proceeds, this is only an approximation. Cracks, changes in concrete composition and moisture levels with depth will also lead to deviation from the perfect diffusion equation. Integration of equation 3.3 gives a square root law that can be used to estimate the movement of the carbonation front. The calculation of diffusion rates is discussed in more detail in Chapter 8.

Empirically, a number of equations have been used to link carbonation rates, concrete quality and environment. Table 3.1 summarizes some of those equations and shows the factors that have been included. Generally there is a t dependence. As discussed above the other factors are exposure, water/cement ratio, strength and CaO content (functions of cement type and its alkali content). For example if we consider the basic equation:

$$d = At^{0.5}$$

and carbonation depths are 16 mm in 16 years in typical poor concrete and 4 mm in 20 years in a good concrete, we would expect the diffusion coefficient A to range from 0.25 to 1.0 mm yr$^{-\frac{1}{2}}$. In bridges it is common to see negligible carbonation after 20 years or more. On buildings it can range from negligible to 60 mm or more. This is a function of the higher cover, lower water cement ratios, better specification and construction required to produce the high strength concrete required in a bridge. However, high strength is not always synonymous with a low diffusion constant. Permeability measurements (Section 4.12) should always be made on structures in harsh environments rather than just relying on the concrete strength as a measure of its durability.

3.2 CHLORIDE ATTACK

3.2.1 Sources of chlorides

Chlorides can come from several sources. They can be cast into the concrete or they can diffuse in from the outside. Chlorides cast into concrete can be due to:

- deliberate addition of chloride set accelerators (calcium chloride, $CaCl_2$, was widely used until the mid-1970s);
- use of sea water in the mix;
- contaminated aggregates (usually sea dredged aggregates which were unwashed or inadequately washed).

Table 3.1 A selection of carbonation depth equations

Equation	Parameters
$d = At^n$	d = carbonation depth t = time in years A = diffusion coefficient n = exponent (approximately $\frac{1}{2}$)
$d = ABCt^{0.5}$	$A = 1.0$ for external exposure $B = 0.07$ to 1.0 depending on surface finish $C = R(wc - 0.25)/(0.3(1.15 + 3wc))^{1/2}$ for water cement ratio $(wc) \geqslant 0.6$ $C = 0.37R(4.6wc - 1.7)$ for $wc < 0.6$ R = coefficient of neutralization, a function of mix design and additives
$d = A(Bwc - c)t^{0.5}$	A is a function of curing B and C are a function of fly ash used
$d = 0.43(wc - 0.4)(12(t - 1))^{0.5} + 0.1$	28 day cured
$d = 0.53(wc - 0.3)(12t)^{0.5} + 0.2$	uncured
$d = (2.6(wc - 0.3)^2 + 0.16)t^{0.5}$	sheltered
$d = ((wc - 0.3)^2 + 0.07)t^{0.5}$	unsheltered
$d = 10.3e^{-0.123f28}$ at 3 years	unsheltered fX = strength at day X
$d = 3.4e^{-0.34f28}$ at 3 years	sheltered
$d = 680(f28 + 25)^{-1.5} - 0.6$ at 2 years	
$d = A + B/f28^{0.5} + c/(CaO - 46)^{0.5}$	CaO is alkali content expressed as CaO
$d = (0.508/f35^{0.5} - 0.047)(365t)^{0.5}$	
$d = 0.846(10wc/(10f7)^{0.5} - 0.193 - 0.076wc)(12t)^{0.5} - 0.95$	
$d = A(T - t_i)t^{0.75}C_1/C_2)^{0.5}$	t_i = induction time T = temperature in °K C_1 = CO_2 concentration C_2 = CO_2 bound by concrete

Source: Parrott (1987).

Chlorides can diffuse into concrete as a result of:

- sea salt spray and direct sea water wetting;
- deicing salts;
- use of chemicals (structures used for salt storage, brine tanks, aquaria, etc.).

Much of our discussion will centre on the diffusion of chlorides into concrete as that is the major problem in most parts of the world either due to marine salt spray or use of deicing salts. However, the cast-in chlorides must not be overlooked, especially when they are part of the problem. A low level of chloride cast in can lead to rapid onset of corrosion if further chlorides become available from the environment. This often happens in marine conditions where seawater contaminates the original concrete mix and then diffuses into the hardened concrete.

3.2.2 Chloride transport through concrete

Like carbonation, the rate of chloride ingress is often approximated to the laws of diffusion. There are further complications here. The initial mechanism appears to be suction, especially when the surface is dry. Salt water is rapidly absorbed by dry concrete. There is then some capillary movement of the salt-laden water through the pores followed by 'true' diffusion. There are other opposing mechanisms that slow the chlorides down. These include chemical reaction to form chloroaluminates and adsorption on to the pore surfaces.

The other problem with trying to predict the chloride penetration rate is defining the initial concentration, as chloride diffusion produces a concentration gradient not a 'front'. In other words we can use the square root relationship for the carbonation front as the concrete either is or is not carbonated, but we cannot use it so easily for chlorides as there is no chloride 'front', but a concentration profile in the concrete. A typical chloride profile is shown in Figure 3.2. This particular profile is a very convincing fit to a diffusion curve but shows no error bars. Many profiles show far more scatter. The calculation of chloride diffusion rates is discussed more fully in Chapter 8.

3.2.3 Chloride attack mechanism

In Chapter 2 we discussed the corrosion of steel in concrete and the effectiveness of the alkalinity in the concrete pores producing a passive layer of protective oxide on the steel surface which stops corrosion. In the previous section we observed that alkalinity in the concrete pores is neutralized by carbonation. The depassivation mechanism for chloride attack is somewhat different. The chloride ion attacks the passive layer but, unlike carbonation, there is no overall drop in pH. Chlorides act as catalysts to corrosion when there is sufficient concentration at the rebar surface to break down the passive layer. They are not consumed in the process but help to break down the passive layer of oxide on the steel and allow the corrosion process to proceed quickly. This is illustrated in

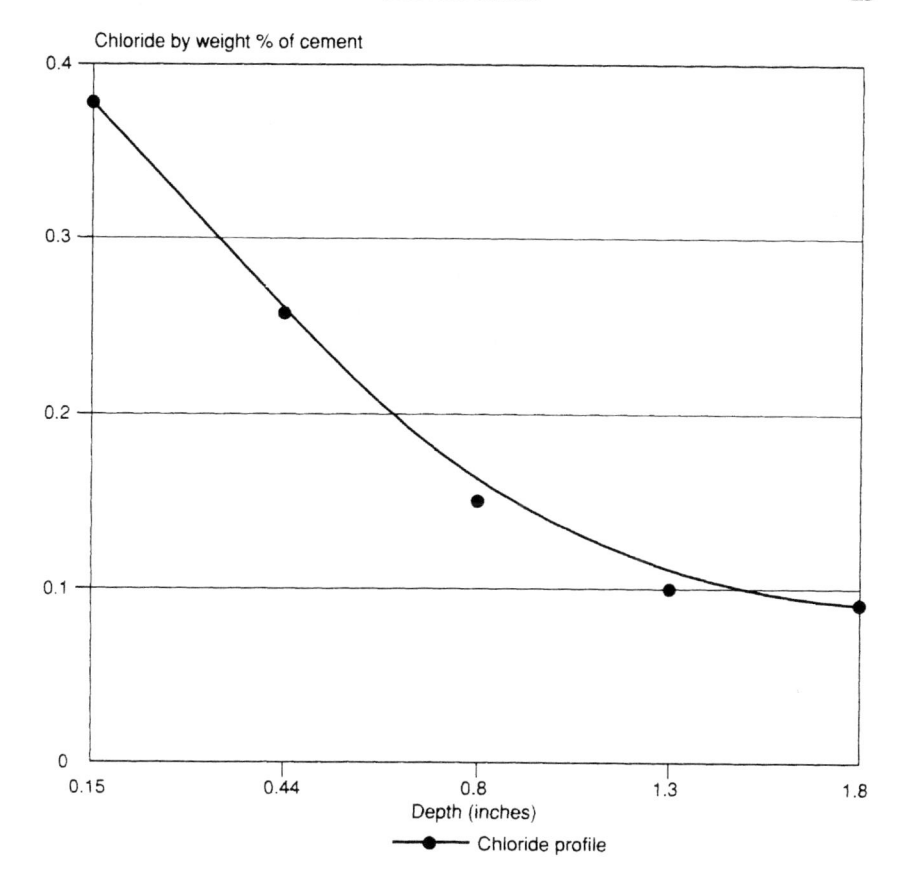

Figure 3.2 Chloride profiles of marine bridge substructure (Yaquina Bay bridge soffit).

Figure 3.3. This makes chloride attack difficult to remedy as chlorides are hard to eliminate.

Obviously a few chloride ions in the pore water will not break down the passive layer, especially if it is effectively re-establishing itself when damaged, as discussed in Chapter 2.

There is a 'chloride threshold' for corrosion given in terms of the chloride/hydroxyl ratio. It has been measured in laboratory tests with calcium hydroxide solutions. When the chloride concentration exceeds 0.6 of the hydroxyl concentration, corrosion is observed (Hausmann 1967). This approximates to a concentration of 0.4% chloride by weight of cement if chlorides are cast into concrete and 0.2% if they diffuse in.

Concrete

$$FeCl_2 \rightarrow Fe^{2+} + 2Cl^-$$

Passive layer | gamma Fe_2O_3

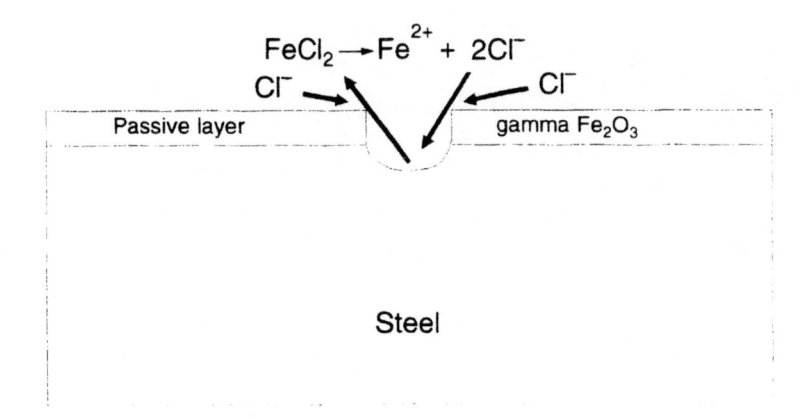

Steel

Figure 3.3 The breakdown of the passive layer and 'recycling' chlorides.

In the USA a commonly quoted threshold is 1 lb chloride per cubic yard of concrete. Although these figures are based on experimental evidence, the actual values are a function of practical observations of real structures.

All these thresholds are approximations because:

1. Concrete pH varies with the type of cement and the concrete mix. A tiny pH change represents a massive change in hydroxyl ion (OH^-) concentration and therefore (theoretically) the threshold moves radically with pH.
2. Chlorides can be bound chemically (by aluminates in the concrete) and physically (by adsorption on the pore walls). This removes them (temporarily or permanently) from the corrosion reaction. Sulphate resisting cements have low aluminate (C_3A) content which leads to more rapid diffusion and lower chloride thresholds.
3. In very dry concrete corrosion may not occur even at very high Cl^- concentration as the water is missing from the corrosion reaction, as discussed in Chapter 2.
4. In sealed or polymer impregnated concrete, corrosion may not occur even at a very high Cl^- concentration if no oxygen or moisture is present to fuel the corrosion reaction.
5. Corrosion can be suppressed when there is total water saturation due to oxygen starvation, but if some oxygen gets in, then the pitting corrosion described in Chapter 2 can occur.

Therefore corrosion can be observed at a threshold level of 0.2% chloride by weight of cement if the concrete quality is poor and there are water and oxygen available. In different circumstances no corrosion may be seen at 1.0% chloride or more if oxygen and water are excluded. If the concrete is very dry or totally saturated (as in (3) or (5) above) then a change in conditions may lead to rapid corrosion.

3.2.4 Macrocell formation

As stated in the previous chapter, corrosion proceeds by the formation of anodes and cathodes (Figures 2.1 and 2.2). In the case of chloride attack they are often well separated with areas of rusting separated by areas of 'clean' steel (illustrated in Figure 2.3). This is known as the macrocell phenomenon. Chloride induced corrosion is particularly prone to macrocell formation as a high level of water is usually present to carry the chloride into the concrete and because chlorides in concrete are hygroscopic (i.e. they absorb and retain moisture). The presence of water in the pores increases the electrical conductivity of the concrete. The higher conductivity allows the separation of anode and cathode as the ions can move through the water filled (or water lined) pores.

The separation of corroded areas does not necessarily represent the distribution of chlorides along the rebar. The anodic and cathodic reactions are separated with large cathodic areas supporting small, concentrated anodic areas.

For carbonation to occur the concrete is generally drier (otherwise the CO_2 does not penetrate far). Corrosion is therefore on a 'microcell' level with apparently continuous corrosion observed along the reinforcing.

Stray currents from cathodic protection systems on other objects (such as a nearby pipeline or DC electric railway lines) can also cause corrosion by accelerating the anodic reaction and suppressing the cathodic reaction. In anaerobic conditions, bacterial attack has also been seen. These causes all give rise to the same corrosion mechanism discussed above, but may need specialist investigation and treatment.

3.3 CORROSION DAMAGE

In most industries corrosion is a concern because of wastage of metal leading to structural damage such as a collapse, perforation of containers and pipes, etc. Most problems with corrosion of steel in concrete are not due to loss of steel but the growth of the oxide. This leads to cracking and spalling of the concrete cover.

Structural collapses of reinforced concrete structures due to corrosion are rare. The author knows of two multistorey parking structures in

North America which have collapsed due to deicing salt induced corrosion. A post-tensioned concrete bridge collapsed in Wales due to deicing salt attack on the strands (Woodward and Williams, 1988), and so did one in Belgium. Concrete damage would usually have to be well advanced before a reinforced concrete structure is at risk.

Particular problems arise when the corrosion product is the black rust described in Chapter 2 and in prestressed, post-tensioned structures where corrosion is difficult to detect as the tendons are enclosed in ducts. Tendon failure can be catastrophic as tendons are loaded to 50% or more of their ultimate tensile strength and modest section loss leads to failure under load. The particular problems of assessing prestressed post-tensioned concrete structures are addressed in Section 4.15.

The most common problem caused by corrosion is spalling of concrete cover. A man was killed in New York City by a slab of concrete which spalled off a bridge substructure due to deicing salts, and a vehicle was badly damaged in Michigan in a similar incident. Special metal canopies have been built around the lower floors of high rise buildings where corrosion has led to risk of falling concrete. This enables the investigators to collect the fallen concrete at regular intervals and weigh it. In that way they can determine whether the corrosion rate is stable, increasing or decreasing.

The important factors in corrosion of steel in concrete compared with most other corrosion problems are the volume of oxide and where it is formed. A dense oxide formed at high temperatures (such as in a power station boiler) usually has twice the volume of the steel consumed. In most aqueous environments the excess volume of oxide is transported away and deposits on open surfaces within the structure. For steel in concrete two factors predominate. The main problem is that the pore water is static and there is no transport mechanism to move the oxide away from the steel surface. This means that all the oxide is deposited at the metal/oxide interface. The second problem is that the oxide is not dense. It is very porous and takes up a very large volume, up to ten times that of the steel consumed. This is illustrated in Figure 3.4.

The thermodynamics of corrosion, coupled with the low tensile strength of concrete, means that the formation of oxide breaks up the concrete. It has been suggested that less than 100 μm of steel section loss are needed to start cracking and spalling the concrete. The actual amount needed will depend upon the geometry in terms of cover, proximity to corners, rebar spacing, bar diameter and the rate of corrosion. There is further discussion of the rate and amount of corrosion leading to damage in Section 4.11 on corrosion rate measurement.

Corners tend to crack first on corroding reinforced concrete struc-

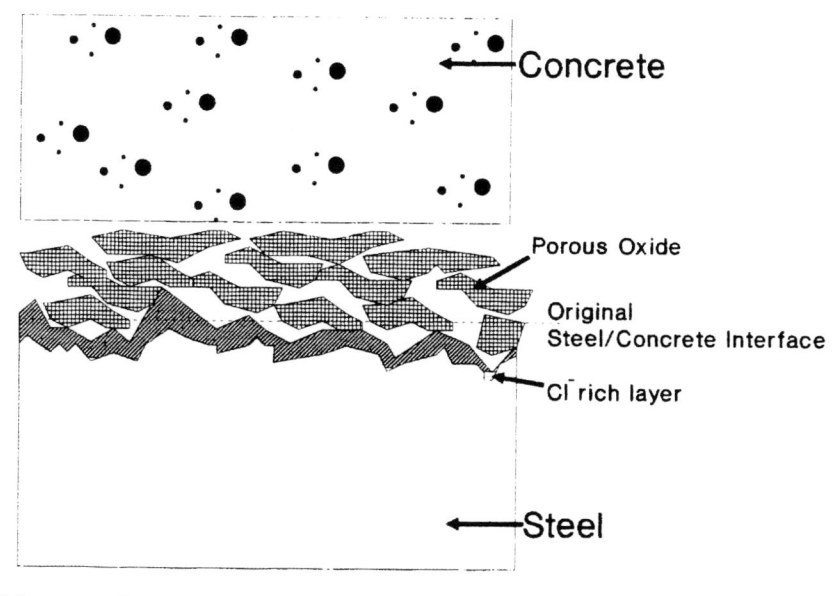

Figure 3.4 Rust growth forcing steel and concrete apart.

tures. This is because the oxygen, water, chlorides and carbon dioxide have two faces as pathways to the steel. Delaminations occur as corrosion proceeds on neighbouring rebars and the horizontal cracks join up as shown in Figure 3.5.

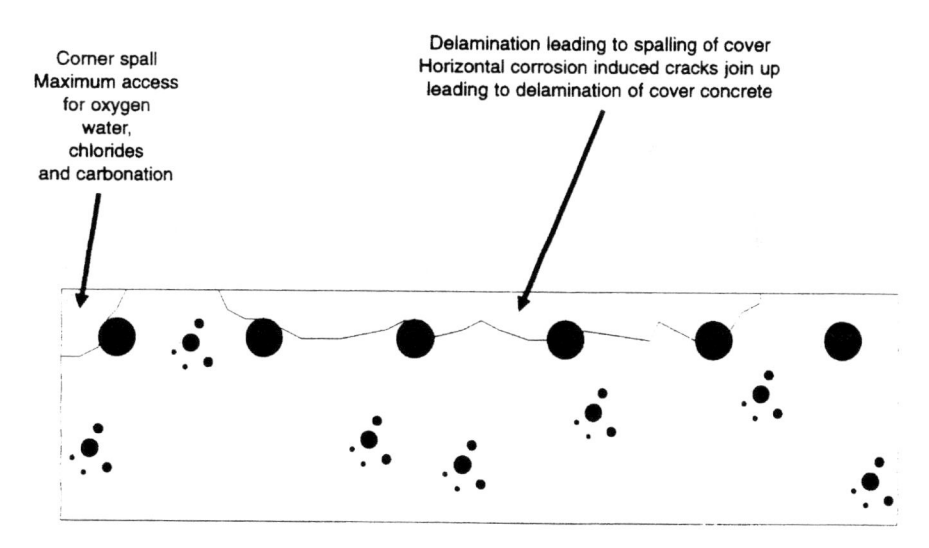

Figure 3.5 Corrosion induced cracking and spalling.

3.4 VERTICAL CRACKS, HORIZONTAL CRACKS AND CORROSION

The importance of vertical cracks in accelerating corrosion by allowing access of corrosion agents to the steel surface has been widely discussed. If reinforcing steel is doing its job in areas of tension in the structure, small cracks will occur as the tensile load exceeds the tensile strength of the steel. Most of these are small cracks (less than 0.5 mm) intersecting the reinforcing steel at right angles. They do not lead to corrosion of the steel as any local ingress of chlorides, moisture and carbonation is limited and contained by the local alkalinity. Obviously there is a limit to this 'self-healing' ability. If large cracks stay open (greater than 0.5 mm), then corrosion can be accelerated. Such cracks may be due to plastic shrinkage, thermal expansion or other reasons. The relationship between cracks in concrete and reinforcement corrosion is fully discussed in Concrete Society (1995) Technical Report No. 44.

Corrosion causes horizontal cracking along the plane of the rebar and the corner cracking around the end rebar. This leads to the loss of concrete cover as shown in Figure 3.5. This is the main consequence of reinforcement corrosion with its subsequent risk of falling concrete and unacceptable appearance.

3.5 THE SYNERGISTIC RELATIONSHIP BETWEEN CHLORIDE AND CARBONATION ATTACK, CHLORIDE BINDING AND RELEASE

We have already discussed the fact that chlorides can be bound by the concrete. One of the constituents of cement paste is C_3A, a complex inorganic aluminium salt. This reacts with chloride to form chloroaluminates. This removes the chloride from availability in the pore water to cause corrosion. This binding process is strongest for chlorides cast into concrete, and is why it was considered acceptable to use sea water to make concrete for many years.

The extent of binding and its effectiveness is not well understood. However, it is known that a reduction in pH as caused by carbonation will break down the chloroaluminates. This leads to a 'wave' of chlorides moving in front of the carbonation front. Consequently structures with chlorides in them that carbonate are more susceptible to corrosion than those with only one source of problem.

REFERENCES

Concrete Society (1995) *The Relevance of Cracking in Concrete to Corrosion of Reinforcement*, Technical Report No. 44, The Concrete Society, Slough, UK.

References

Hausmann, D.A. (1967) 'Steel corrosion in concrete: how does it occur?', *Materials Protection*, **6**, 19–23.

Parrott, L.J. (1987) *A Review of Carbonation in Reinforced Concrete*, a review carried out by the Cement and Concrete Association under a BRE contract.

Woodward, R.J. and Williams, F.W. (1988) 'Collapse of Ynes-y-Gwas bridge, West Glamorgan', *Proceedings of the Institution of Civil Engineers Part I*, **84**, 635–69.

4
Condition evaluation

We have considered the main mechanisms of corrosion in Chapter 2. We have seen that the chemical process is the same regardless of whether the cause is carbonation or chloride attack, as described in Chapter 3. But if we are to perform an effective repair we must fully understand the cause and extent of damage or we risk wasting resources with an inadequate or unnecessarily expensive repair. This chapter explains how to evaluate the condition of corroding reinforced concrete structures.

A full evaluation is normally a two-stage process. The preliminary survey should characterize the nature of the problem and give guidance in planning a detailed survey. The detailed survey will confirm the cause and quantify the extent of the problem. The Concrete Society Technical Report 26 (Concrete Society, 1984) and the American Concrete Institute Committee 222 Reports (American Concrete Institute, 1990) give excellent reviews of how to conduct the surveys.

It is important to remember that corrosion is not the only deterioration mechanism in reinforced concrete. Alkali–silica reactivity (ASR), freeze–thaw, plastic shrinkage, thermal movement, settlement and other movement can all lead to cracking and spalling of concrete.

Some structures may be prone to unusual chemical attack of the steel or the concrete. For example:

- Storage vessels can contain liquids that will attack aggregates, cement paste or the steel.
- Carbonates in water can attack concrete pipelines and underground structures.

However, we will concentrate on corrosion of atmospherically exposed reinforced concrete structures and elements.

A condition evaluation as described here is not a structural survey. A structural engineer must be consulted if there are concerns about the capacity of the structure either because of corrosion damage or for any other reasons. Any excessive deflection of structural elements, misalignment, impact damage, excessive cracking, loss of concrete or loss of steel section will require a structural evaluation before repairing corrosion damage.

4.1 PRELIMINARY SURVEY

This normally involves a visual inspection, probing of cracks and spalls to see their extent, reinforcement cover measurement, possibly a few carbonation measurements, half cell measurements and the taking of samples (sometimes taking broken pieces of concrete rather than coring) for laboratory testing. Particular attention must be paid to safety and structural integrity from concrete spalling or steel section loss. Other causes of concrete cracking (e.g. ASR, freeze–thaw, thermal movement, structural movement, impact damage) must not be overlooked at this stage.

4.2 DETAILED SURVEY

The purpose of a detailed survey is to define the extent and severity of deterioration as accurately as possible. We will need to know how much damage has been done and what has caused the damage. Quantities for repair tenders will probably be based on the results of this survey, so a full survey of all affected elements may be required. Alternatively a full visual and hammer (delamination) survey may be required, with detailed measurements of half cell potentials, chloride contents, carbonation depth, cover, etc. at a few representative locations. It is usual to produce pro formas for noting down all deterioration, as well as readings and samples taken so that everything can be tied together in the analysis. The weather conditions are also recorded as these can affect some readings. An example of a report drawing of a condition evaluation is given in Figure 4.1.

4.3 AVAILABLE TECHNIQUES

The following sections explain the available techniques, their advantages and limitations and the resources needed to employ them. If all the techniques available were used to survey a corroding reinforced concrete structure thoroughly then huge resources could be used before anything was done to stop the problem. However, expensive repairs can be useless if the problem is not properly diagnosed. So what techniques should be used and what are their limitations and capabilities?

Table 4.1 lists most of the techniques that are or can be used for

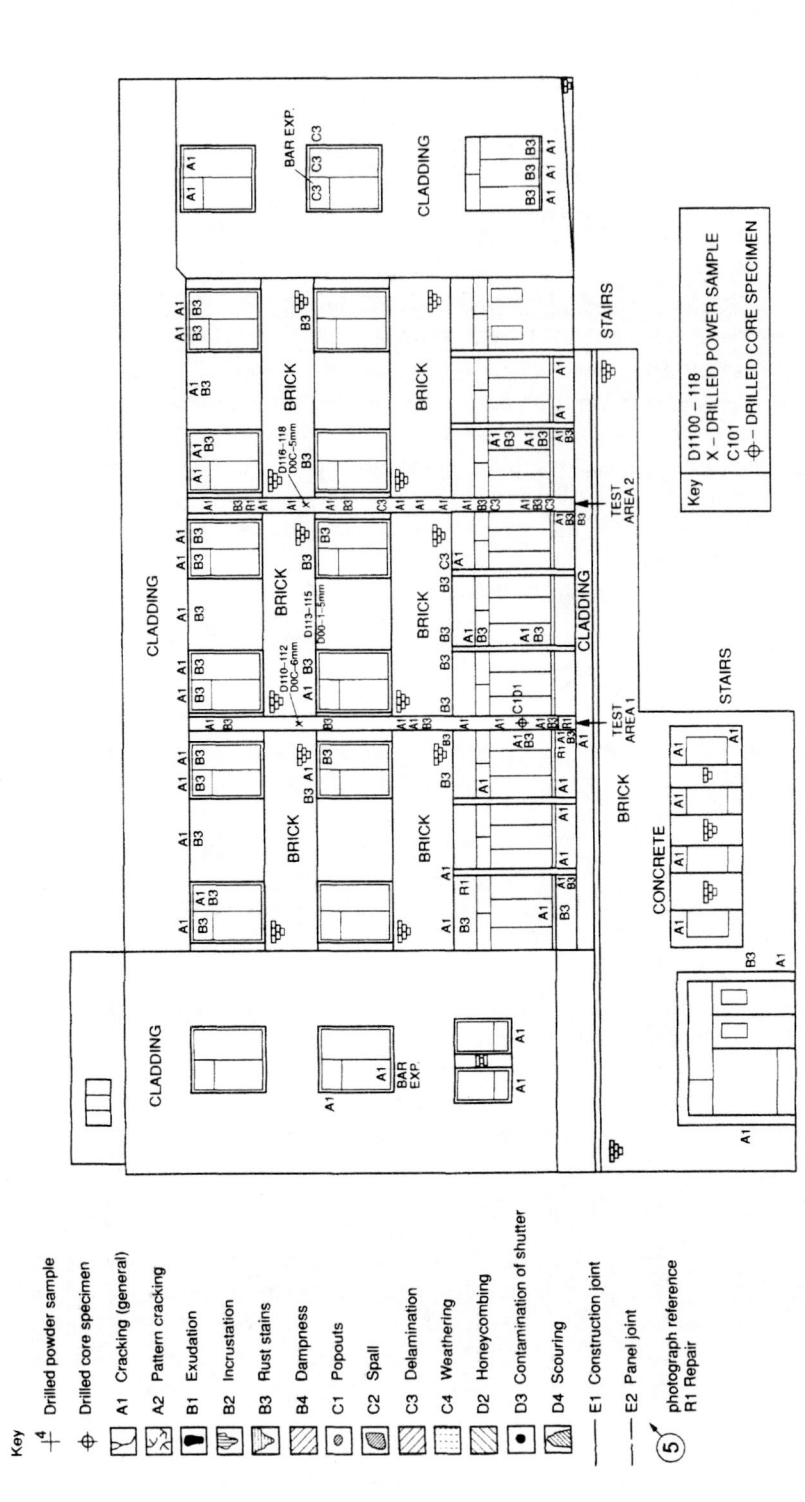

Figure 4.1 Typical visual survey of a RC framed building suffering sea salt spray induced corrosion on top of cast-in chlorides.

Table 4.1 Methods for condition surveying

Method	Detects	Use	Approximate speed
Visual	Surface defects	General	1 m^2 s^{-1}
Hammer/chain	Delaminations	General	0.1 m^2 s^{-1}
Cover meter	Rebar depth and size	General	1 reading in 5 min
Phenolphthalein	Carbonated depth	General	1 reading in 5 min
Chloride content	Chloride corrosion	General	Core in 10 min or drillings in 2 min + lab or special site analysis
Half cell	Corrosion risk	General/specialist	1 reading in 5 s
Linear polarization	Corrosion rate	General/specialist	1 reading in 5–30 min depending on equipment used
Resistivity	Concrete resistivity/corrosion risk	General/specialist	1 reading in 20 s
Permeability	Diffusion rate	General/specialist	1 reading in 5 min or core + lab
Impact/ultrasonics	Defects in concrete	Specialist	1 reading in 2 min
Petrography	Concrete condition, etc.	General	Core + lab
Radar/radiography	Defects, steel location, condition	Specialist	>1 m^2 per second for vehicle system or 1 m^2 in 20 s for hand system + interpreting

condition surveying and have relevance to corrosion or the damage it causes. A minimum requirement would usually be a visual survey, a delamination survey, carbonation and chloride measurements, and cover measurements. A petrographic analysis of the concrete is also usually required. These will be done in one or two representative or more severely damaged areas. These measurements will tell the engineer the cause of corrosion, the extent to which chlorides or carbonation have depassivated the steel and the extent of existing damage.

Half cell surveys are increasingly popular as a rapid way of showing how corrosion is spreading ahead of the damage it causes. The other techniques are used for specific requirements as described below.

4.4 VISUAL INSPECTION

The visual inspection is the first step in any investigation. It may start as a casual 'look over' that spots a problem and end up as a rigorous logging of every defect seen on the concrete surface (Concrete Society, 1984).

4.4.1 Property to be measured

The aim of the visual survey is to give a first indication of what is wrong and how extensive the damage is. If concrete is spalling off then that can be used as a measure of extent of damage. Weighing the concrete that spalls off over set periods can be used as a direct measure of the deterioration rate.

4.4.2 Equipment and use

The main equipment is obviously the human eye and brain, aided with a notebook, pro forma or hand-held computer and a camera. Binoculars may be necessary, but close inspection is better if access can be arranged. A systematic visual survey will be planned in advance. Many companies that carry out condition surveys will have standardized systems for indicating the nature and extent of defects. These are used in conjunction with customized pro formas for each element or face of the structure. It is normal to record date, time and weather conditions when doing the survey, also noting visual observations such as water or salt run down and damp areas. Examples are given in Figures 4.1 and 4.2.

4.4.3 Interpretation

Interpretation is usually based on the knowledge and experience of the engineer or technician conducting the survey. The Strategic Highway Research Program (SHRP) has produced an expert system, HWYCON (Kaetzel *et al.*, 1994). This guides the less experienced engineer or technician through the different types of defects seen on concrete highway pavements and structures, including alkali–silica reaction, freeze–thaw damage and corrosion.

4.4.4 Limitations

The main limitation is the skill of the operative. Some defects can be mistaken for others. When corrosion is suspected, it must be understood that rust staining can come from iron bearing aggregates rather than from corroding reinforcement. Different types of cracking can be attributed to different causes. The recognition of ASR is discussed in the SHRP manual on ASR (Stark, 1991) and in HWYCON (Kaetzel

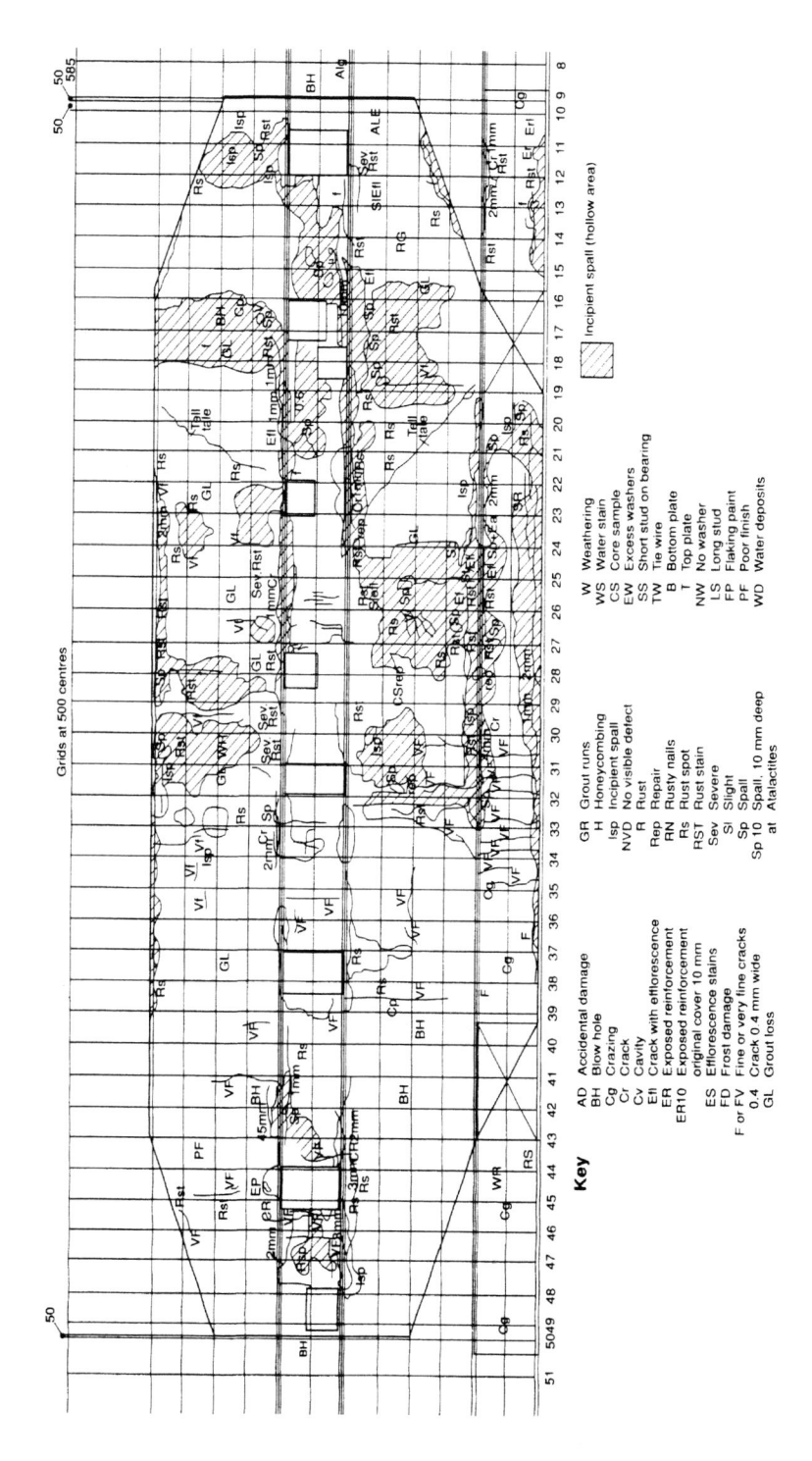

Figure 4.2 Visual delamination survey on a cross beam on a motorway bridge in the UK suffering from deicing salt ingress (developed elevation of beam viewed in the direction of increasing bent numbers).

et al., 1994). Visual surveys must always be followed up by testing to confirm the source and cause of deterioration.

4.5 DELAMINATION

As corrosion proceeds, the corrosion product formed takes up a larger volume than the steel consumed. This builds up tensile stresses around the rebars. A layer of corroding rebars will often cause a planar fracture at rebar depth, before the concrete spalls, as shown earlier in Figures 3.4 and 3.5. This can be detected at the surface by various means from hitting the surface with a hammer and listening for a hollow sound to sophisticated techniques using radar, infrared, sonic and ultrasonic equipment.

4.5.1 Property to be measured

The aim is to measure the amount of cracking between the rebars before it becomes apparent at the surface. It should be noted that this can be a very dynamic situation. Figure 3.4 shows the way that cracks propagate between corroding bars. The horizontal cracks are detected by a hollow sound when the surface is hit with a hammer or a chain dragged across the surface. It can be detected from the effect on its physical properties associated either with the presence of a layer of air in the concrete or the phase change from concrete to air to concrete, when subjected to radiation or ultrasound.

4.5.2 Equipment and use

The hammer survey or chain drag (on decks) is usually quicker, cheaper and more accurate than the other alternatives such as radar, ultrasonics or infrared thermography. However, these techniques do have their uses, for instance in large-scale surveys of bridge decks (radar and infrared thermography), of waterproof membranes or other concrete defects (ultrasonics and radar). They are discussed later in this section.

The delamination survey with a hammer is often conducted at the same time as the visual survey. Hollow sounding areas are marked directly on to the surface of the structure with a suitable permanent or temporary marker and recorded on the visual survey pro forma.

Correctly tuned infrared cameras can be used to detect the temperature difference between solid and delaminated concrete. This is best done when the concrete is warming up or cooling down as the delaminated concrete heats and cools faster. This means that the technique's sensitivity depends upon the weather conditions and the orientation of the face being surveyed. Infrared thermography tends to work best on bridge decks in the early morning or late evening of clear days. The best systems incorporate a visible light camera for joint recording of visual and IR image so that the location of defects is recorded simultaneously.

Radar measures changes in the dielectric constants associated with the concrete/air phase change. However, the radar also senses the dielectric changes at the steel–concrete interface, the presence of water and, to a small extent, chlorides. This makes interpretation of radar images a difficult process. In North America the main use of radar and infrared has been for bridge deck surveys with vehicle mounted systems. In Europe and the UK hand-held systems have been used for surveys of building and other structures. The reader is recommended to review the literature for further information (Cady and Gannon, 1992; Bungey, 1993; Titman, 1993). There is further discussion of this topic in Section 4.12.3.

Radar is used in North America for surveying bridge decks using truck mounted rapid data acquisition. Radar is not accurate in defining the size and location of individual delaminations but can be used for generalized condition surveys or comparative ranking of damaged decks (Alongi et al., 1993). Radar and infrared thermography have been used in combination in North America. This increases the accuracy of the measurement of delaminations and other defects but at the expense of doubling equipment and interpretation costs. The problem of needing the right weather conditions for the infrared thermograms has led to a decline in the popularity of this approach.

4.5.3 Interpretation

The interpretation of radar and infrared is a specialist process usually carried out by the companies who have the equipment and are hired to conduct such surveys.

With a hammer or chain drag survey the experience of the operative is vital. A skilled technician who is experienced in carrying out delamination surveys will often produce better and more consistent results than the more qualified but less experienced engineer.

4.5.4 Limitations

The trapping of water within cracks, deep delaminations (where bars are deep within the structure) and heavy traffic noise can complicate the accurate measurement of delaminations for hammer techniques. Water and deep delaminations also cause problems for radar and infrared thermography.

It is common during concrete repairs for the amount of delamination to be far more extensive than delamination surveys indicate. This is partly due to the inaccuracy of the techniques available but also because of the time between survey and repair. Once corrosion has started, delaminations can initiate and grow rapidly. An underestimate of 40% or more is not unusual and should be borne in mind when budgeting for repairs.

Radar is reasonably accurate in predicting the amount of damage on a bridge deck, but not the precise location of the damage. The problems with infrared thermography are getting the right weather conditions to carry out a useful survey.

4.6 COVER

Cover measurement is carried out on new structures to check that adequate cover has been provided to the steel according to the specifications. It is also carried out when corrosion is observed because low cover will increase the corrosion rate both by allowing the agents of corrosion (chlorides and carbonation) more rapid access to the steel, and also allowing more rapid access of the 'fuels' for corrosion, moisture and oxygen. A cover survey will help explain why the

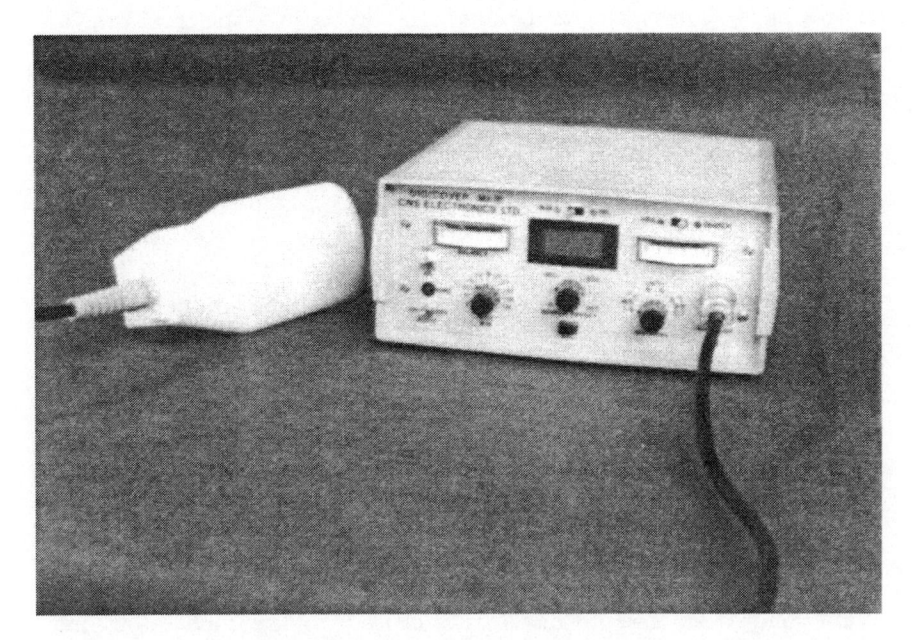

Figure 4.3 The Digicover Mark III cover meter. Courtesy of CNS Electronics.

structure is corroding and show which areas are most susceptible to corrosion due to low cover.

4.6.1 Property to be measured

A cover survey requires the location of the rebars to be measured in three dimensions, i.e. their position with regard to each other and the plane of the surface (X, Y) and depth from the surface (Z). If construction drawings are not available then it may be necessary to measure the rebar diameter as well as its location.

4.6.2 Equipment and use

Magnetic cover meters are now available with logging capabilities and digital outputs. A spacer can be used to estimate rebar diameter. Other alternatives such as radiography can be used to survey bridges or other structures but this is rarely cost effective (Cady and Gannon, 1992; Bungey, 1993). Cover meters are surprisingly difficult to use. They are slow, and deep cover and closely spaced bars affect the readings. A typical device is shown in Figure 4.3. An alternating magnetic field is used to detect the presence of magnetic materials such as steel rebars.

4.6.3 Interpretation

One of the few standards for cover meters is BS1881, Part 204. This refers to the measurement on a single rebar and is more concerned with the meter than taking a cover survey. Alldred (1993) discusses cover meter accuracy when several rebars are close together and suggests that different types of head are more accurate in different conditions. The smaller heads are better for resolving congested rebars.

4.6.4 Limitations

The main problem with cover measurements is the congestion of rebars giving misleading information (Alldred, 1993). Iron bearing aggregates can lead to misleading readings as they will influence the magnetic field. Different steels also have different magnetic properties (at the extreme end, austenitic stainless steels are non-magnetic). Most cover meters have calibrations for different reinforcing steel types. The devices are slow and are not very accurate in the field, as anyone who has tried to use one to locate steel and excavate the steel will know. You often miss the steel that is indicated when excavating to find it to make electrical connections or visually examine it. It is always advisable to check cover meters by excavating and exposing one or more rebars.

4.7 HALF CELL POTENTIAL MEASUREMENTS

The electrochemistry of corrosion, cells and half cells was discussed in Section 2.4. The half cell is a simple device. It is a piece of metal in a solution of its own ions (such as copper in copper sulphate, silver in silver chloride, etc.). If we connect it to another metal in a solution of its own ions (such as iron in ferrous hydroxide, $Fe(OH)_2$, see equations 1.1 and 1.3) there will be a potential difference between the two 'half cells'. We have made a battery (or an electrical single cell to be precise). It will generate a voltage because of the different positions of the two metals in the electrochemical series (Table 2.1) and due to the difference in the solutions (Figure 2.6). This is a galvanic cell in that the corrosion and current flow between different metals is known as galvanic action. A second type of cell is a concentration cell that will generate a voltage depending upon differences in the concentration of the solution (strictly the activity), around otherwise similar electrodes.

By using a standard half cell that is in a constant state, and moving it along the concrete surface, we change our full cell by the difference in condition of the steel surface below the moving half cell. If the steel is passive the potential measured is small (zero to −200 mV against a copper/copper sulphate half cell, or even a positive reading); if the passive layer is failing and increasing amounts of steel are dissolving (or if small areas are corroding but the potential is being averaged out with passive area), the potential moves towards −350 mV. At more negative than −350 mV the steel is usually corroding actively. By convention we connect the positive terminal of the voltmeter to the steel and the negative terminal to the half cell. This gives a negative reading.

Very negative potentials can be found in saturated conditions where there is no oxygen to form a passive layer; but with no oxygen there can be no corrosion. This shows the weakness of potential measurements. They measure the thermodynamics of the corrosion, not the rate of corrosion. Corrosion potentials can be misleading. Their interpretation is based on empirical observation, not rigorously accurate scientific theory. The problem is that the potential is not purely a function of the corrosion condition but also of other factors, and that the corrosion condition is not the corrosion rate.

The half cell potential measurement gives an indication of the corrosion risk of the steel. The measurement is linked by empirical comparisons to the probability of corrosion.

The half cell potential is a function of the amount of iron dissolving (corroding). This is a function of the extent to which the steel is depassivated, i.e. the extent of carbonation around the steel or the presence of sufficient chloride to break down the passive layer, and the presence of oxygen to sustain the passive layer. Without oxygen, iron will dissolve but will remain stable in solution as there is no compensating cathodic reaction so the potential may be very negative. However, the corrosion rate will be low. However, there may be a very large (negative) potential against a standard half cell. A detailed description of the half cell and its use is given in Vassie (1991).

4.7.1 Equipment and use

The equipment used is a standard half cell. Silver/silver chloride (Ag/AgCl) and mercury/mercury oxide (Hg/HgO) are recommended. Copper/copper sulphate (CSE) cells are also used but are not recommended because of the maintenance needs, the risk of contamination of the cell, the difficulty of use in all orientations and the leakage of copper sulphate.

It is important to record the equipment used. Different half cells have different 'offsets'. The silver/silver chloride half cell gives potentials that are a function of the chloride concentration in them. This is usually about 130 mV more positive than a copper/saturated copper sulphate electrode. This can be compensated for internally in the logging equipment if used or during reporting if the ASTM criteria are being used (see below).

Standard electrode potentials are given against the 'hydrogen scale'. That is against a standard cell consisting of one atmosphere (strictly unit fugacity) of hydrogen gas in a solution containing one mole (strictly unit activity) of hydrogen ions. The cell itself has a platinum electrode with hydrogen bubbling over it and is in a 1M solution of hydrochloric acid.

It is more usual to measure potentials in the laboratory against a saturated calomel electrode (mercury in saturated mercuric chloride). This cell is recommended for calibrating field half cells. It is possible to adjust the chloride content of a silver/silver chloride cell so that it behaves like a calomel cell. A calomel cell is not usually used in the field because it contains mercury.

A high impedance digital voltmeter is used to collect the data in the simplest configuration. Other options are to use a logging voltmeter (or logger attached to a voltmeter), an array of cells with automatic logging or a half cell linked to a wheel for rapid data collection (Broomfield et al., 1990). Examples are shown in Figures 4.4(a)–(d).

Loggers can be linked to individual half cells or built into the wheel or array systems. They will store readings and position. They will download to computers and some have built in printers to output the data on site.

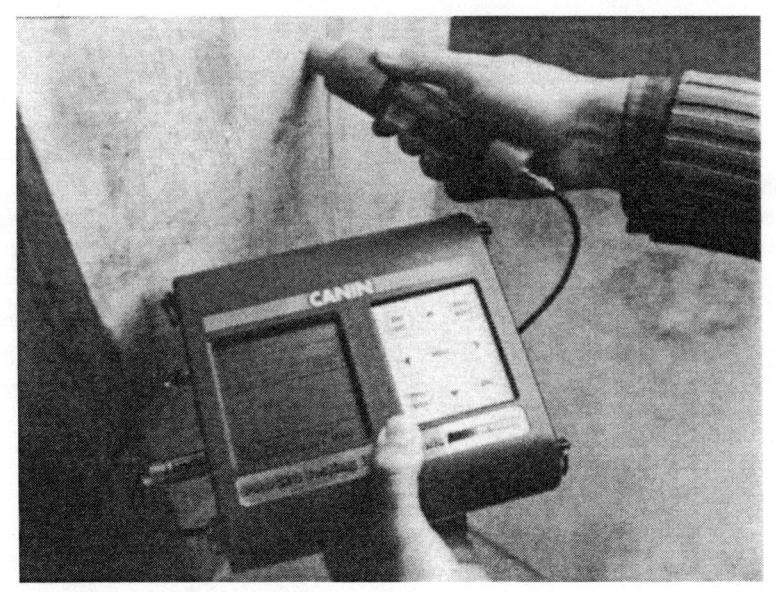

Figure 4.4(a) A half cell and 'Canin' logging voltmeter. Courtesy of Proceq SA and Hammond Concrete Services.

The measurement procedure with a half cell is as follows:

1. Decide on the area of measurement, usually a whole element such as a bridge deck or cross-head beam, or representative areas of an element or the structure, several square metres in size.
2. Use a cover meter to locate the steel and determine rebar spacing.
3. Make an electrical connection to the steel either by exposing it or using already exposed steel.
4. Check that the steel is electrically continuous with a DC resistivity meter between two points that are well separated and on well separated rebars.
5. Mark out a grid. This will typically be 0.2 to 0.5 m^2 but may be smaller, larger or rectangular depending on the steel spacing, the geometry of the element being surveyed and other factors determined by the experience of the investigator. The grid may coincide with the rebar spacing on small surveys, but not usually on larger scans.
6. Check and calibrate the half cell and voltmeter.
7. If necessary wet the whole area to ensure good electrical contact Alternatively wet the immediate area of the measurement. Tap water, soap solution and even saline solutions have been recommended for wetting. The author prefers tap water (for a reading to be made charged ions must flow from the steel to the half cell; the

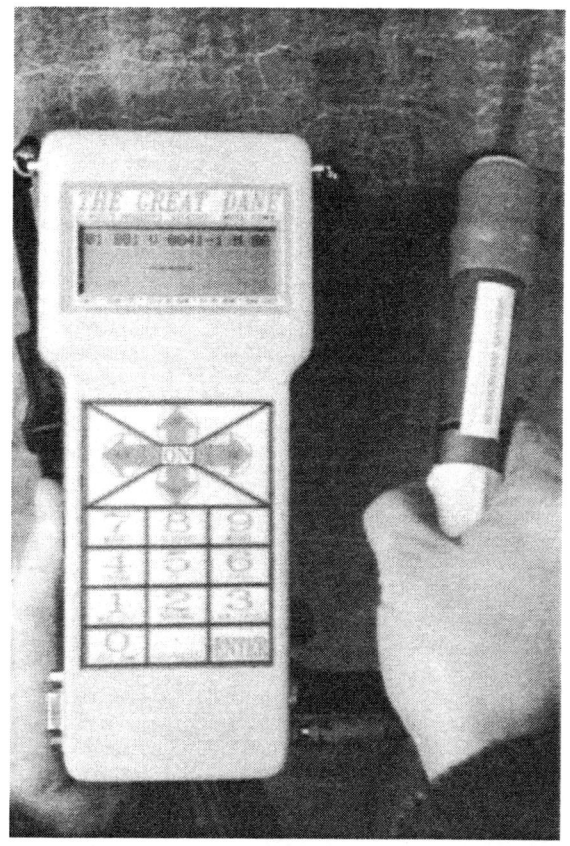

Figure 4.4(b) Half cell potential measurement with the 'Great Dane' logging voltmeter and display. Courtesy of Germann Instruments.

concrete must therefore be damp enough for an ionic path; direct contact to the steel must not occur, the current must flow as ions, not electrons).
8. Take and record the readings. For manual measurements (without automated logging equipment) it is good practice to take two immediately adjacent readings to check that they are within a few millivolts of each other.
9. Examine for anomalies, check most negative reading areas for signs or causes of corrosion.

Data are normally recorded on a plan reflecting the survey grid. The interpretation and presentation are discussed below. The potential map should be drawn up while still on site in order to check that the data are sensible and that apparent 'corrosion hot spots' are investigated as part of the survey.

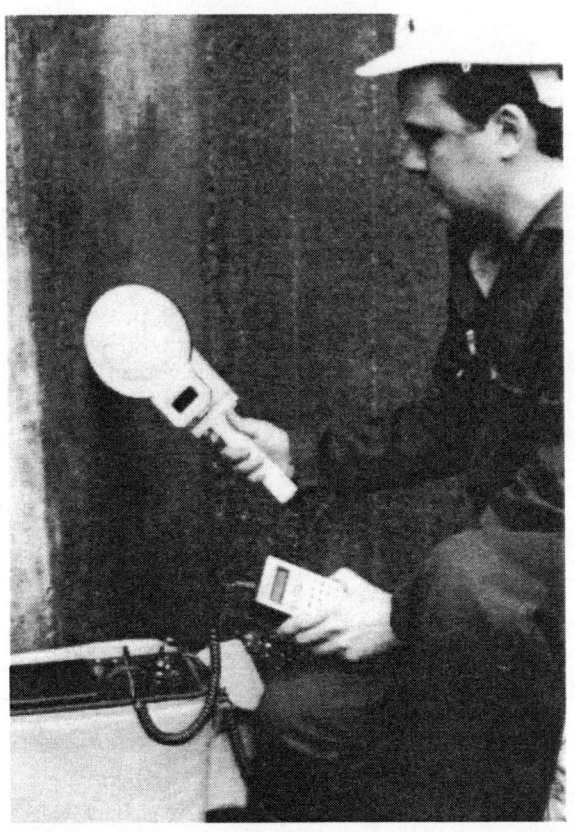

Figure 4.4(c) The potential voltmeter for half cell potential measurements. Courtesy CNS Electronics.

4.7.2 Interpretation and the ASTM criteria

ASTM C867 presents one way of interpreting half cell potentials in the field. ASTM quotes the values against a copper/copper sulphate half cell. However, as noted above, the copper/copper sulphate cell is not recommended and cells should be calibrated against a calomel cell. Therefore the criteria are given in Table 4.2 against a saturated calomel electrode (SCE), the standard hydrogen electrode (SHE) and a routinely used silver/silver chloride electrode. The negative sign is by convention and will depend upon how the leads are connected to the half cell and the rebar from the millivoltmeter.

This interpretation was devised empirically from salt induced corrosion of cast-in-place bridge decks in the USA. Also offsets are seen for different types of concrete (precast or containing cement replacement materials). Figure 4.5 shows the ranges of potentials seen on different structures with different concretes and different conditions.

Half cell potential measurements

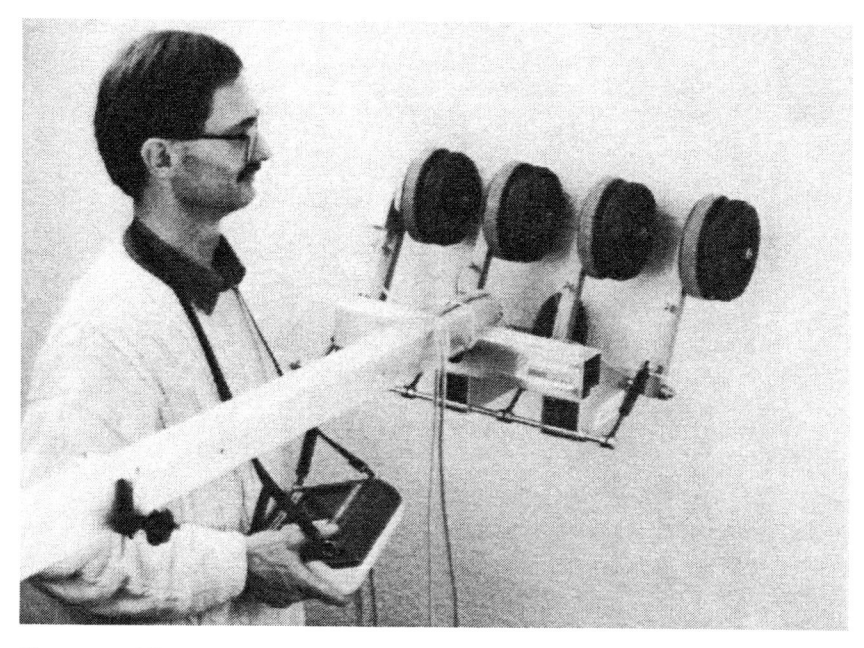

Figure 4.4 (d) The 'Canin' multiple wheeled half cell array. Courtesy of Proceq SA and Hammond Concrete Services.

The major problems with this interpretation occur when there is little oxygen present, especially where the concrete is saturated with water and the potentials can go very negative without corrosion occurring. The wet bases of columns or walls often show more negative potentials

Table 4.2 ASTM criteria for corrosion of steel in concrete for different standard half cells

Copper/copper sulphate	Silver/silver chloride/ 4M KCl	Standard hydrogen electrode	Calomel	Corrosion condition
> −200 mV	> −106 mV	> +116 mV	> −126 mV	Low (10% risk of corrosion)
−200 to −350 mV	−106 to −256 mV	+116 mV to −34 mV	−126 mV to −276 mV	Intermediate corrosion risk
< −350 mV	< −256 mV	< −34 mV	< −276 mV	High (<90% risk of corrosion)
< −500 mV	< −406 mV	< −184 mV	< −426 mV	Severe corrosion

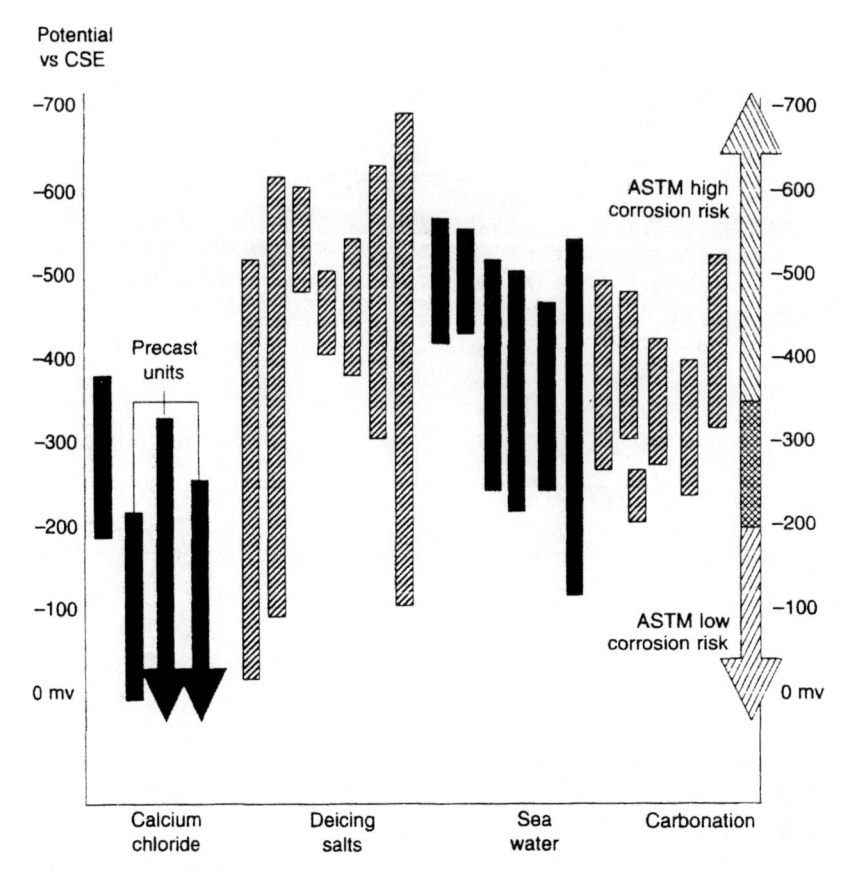

Figure 4.5 Potential ranges (rebar potential vs. a $Cu/CuSO_4$ half cell). From Baker (1986).

regardless of corrosion activity. Very negative potentials have been measured below the water line in marine environments; however, the lack of oxygen will often slow the corrosion rate to negligible levels. Other problems arise in the presence of carbonation.

In carbonated concrete the anodes and cathodes are so close together that a 'mixed potential' (an average of the anode and cathode) is measured. Also carbonated concrete wets and dries quickly as the pores are partly blocked by the calcium carbonate deposits. This means that the resistivity of the concrete will affect the measurement. If the reading is taken with no wetting a very positive reading may be found. If the reading is taken after wetting the measurement may drift more negative for many hours.

A third problem arises due to the existence of the carbonation front. This is a severe change in the chemical environment from pH 12 to pH 8, i.e. a factor of 10^4 difference in the concentrations of the hydroxyl and hydrogen ions. There are similar changes in the calcium and other metal ion concentrations in solution as they precipitate out on carbonation. This can lead to a 'junction potential' superimposed on the corrosion potential giving rise to misleading results.

For carbonated concrete the best method is first to do a potential survey with minimum wetting (if stable potentials can be established). Then wet the surface thoroughly and leave it for at least two hours or until potentials are stable. Resurvey once potentials have stabilized and then look for the most anodic (negative potential) areas. These are most probably active if a potential difference of 150 mV or more exists over a space of 1.0 m or less. A physical investigation is essential to see if there is reasonable correlation between corrosion and anodic areas.

Given the thermodynamic nature of the measurement and our current understanding of potential measurements, that is probably the best that we can do presently on interpretation of half cell potentials. A fuller description of interpretation methods for chloride contamination is given in Vassie (1991). For all its limitations, the half cell is a very powerful diagnostic tool for corrosion investigation. Its main problem is that some people rely on too simplistic interpretation. Half cell potentials are not corrosion rate measurements. Corrosion rate measurement is discussed below.

Stray electrical currents can also influence the readings. These were discussed in Section 2.3. The effect of stray current on half cell potentials can be used as a diagnostic tool where stray current corrosion is suspected in the presence of DC fields. If a half cell is mounted on or in the concrete linked to a logging voltmeter, then any fluctuations in potentials may be linked to the operation of nearby DC equipment, especially if the equipment can be deliberately turned on and off and the potentials fluctuate accordingly.

There has been a tendency to correlate half cell potentials with corrosion rates. The half cell potential is a mixed potential representing anodic and cathodic areas on the rebar. It is not the driving potential in the corrosion cell. Any correlation between potential and corrosion rate is fortuitous and is often due to holding other variables constant in laboratory tests.

There has been some discussion above and in the literature of the 'junction potentials' created by the change in chemical concentrations within the concrete (Bennett and Mitchell, 1992). This effect was severe in a concrete slab subjected to chloride removal, but that may be due to the treatment (discussed in Section 6.10), rather than being a real problem under normal conditions. However, the junction potential may explain the erratic changes in potentials seen in carbonated structures. This is because potentials exist across the carbonation front due to the pH change, and there are concentration changes as carbonated concrete wets and dries quickly because of the lining of the pores by calcium carbonate.

A histogram or cumulative frequency plot will show what proportion of measurements exceed the criteria to show the extent of high corrosion risk (see ASTM C876, 1991). Where the ASTM criteria do not apply they will show the distribution of readings so that high risk areas can be identified.

The best way of interpreting half cell potential data is to expose areas of rebar which show the most negative potentials, intermediate and least negative potentials to correlate corrosion condition with readings. If there are severe potential gradients across the surface then corrosion is likely to be localized with pitting present. Care must be taken in areas of moisture, e.g. puddles that remain on the surface or where moisture comes up from the ground. These may show very negative potentials due to chloride accumulation or to oxygen starvation. Corrosion rate measurements may be required to determine whether the potential is an artefact or due to high corrosion rates.

4.7.3 Half cell potential mapping

A fuller understanding of the corrosion condition is given by drawing a potential map of the area surveyed. This is a printout of the readings where lines are drawn separating the levels of potential, i.e. joining the points of equipotential with 'contour lines'. This shows the high corrosion risk areas and the low corrosion risk areas. A rapid change in potential is seen as a steeper gradient. This indicates a greater risk of corrosion. Figure 4.6 shows a typical half cell potential plot for a bridge beam, which correlates with the visual survey in Figure 4.2. While these isopotential contour plots are not as quantitative as simply following the ASTM criteria such mapping of anodic areas is valid over a wider range of structures and conditions. The distribution of points between the different contour levels is given.

A line of potential measurements can be plotted on a distance vs. potential plot. This will show which points exceed the ASTM criteria and where the steepest potential gradients are and the most anodic areas are with the most negative values. A set of contour plot, three dimensional plot and line scan is shown in Figure 4.7 (from Broomfield et al., 1990).

4.7.4 Cell to cell potentials

If it is not feasible to make direct connections to the reinforcing steel it is possible to obtain comparative potential data by measuring the

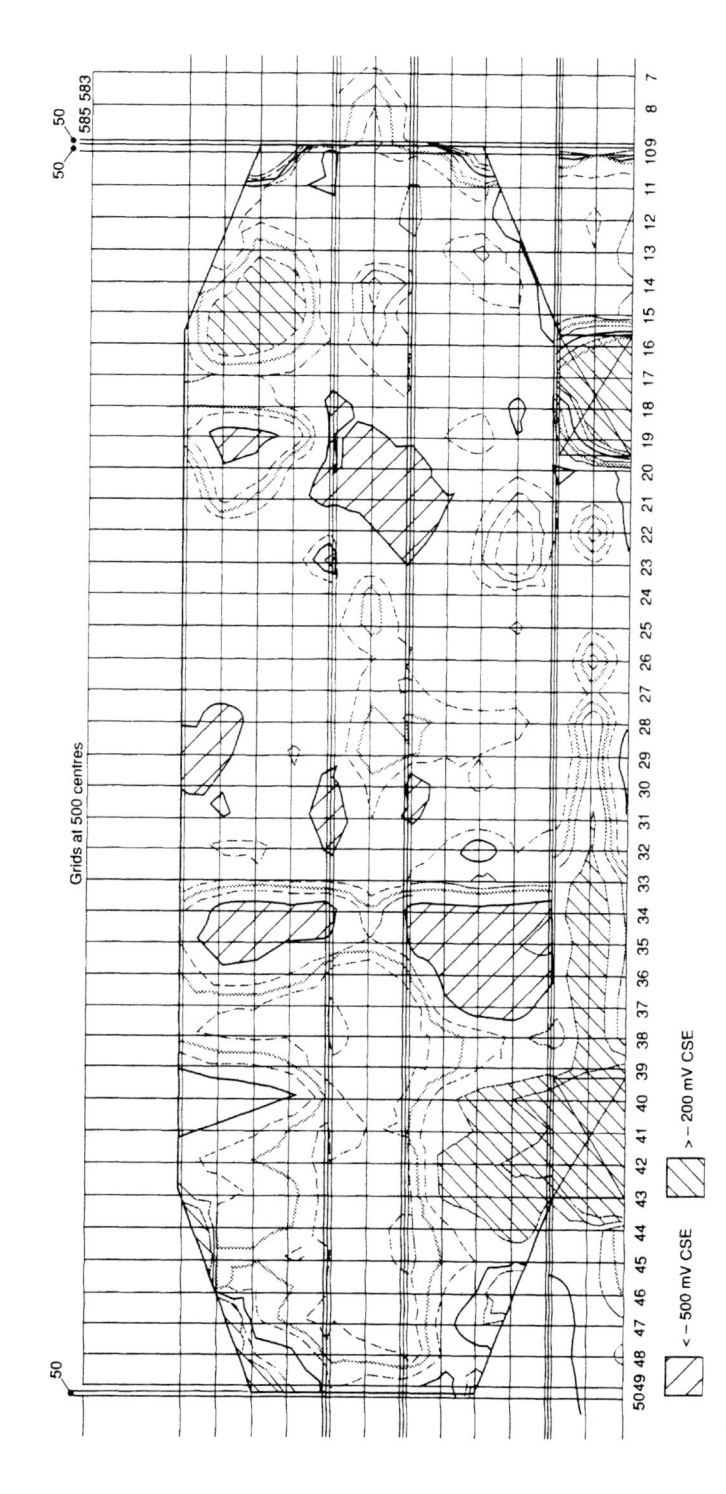

Figure 4.6 Potential survey as per delamination survey (Figure 4.2).

Condition evaluation

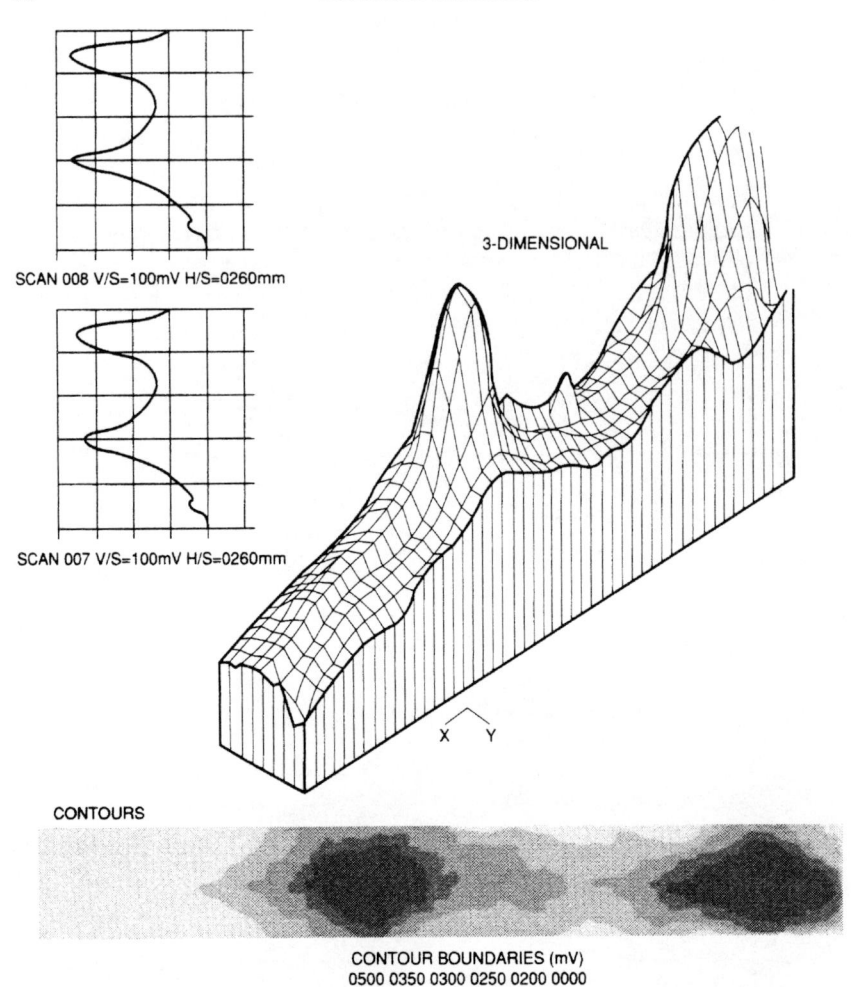

Figure 4.7 Presentation of potential wheel results.

potentials between two half cells, with one kept in a fixed position and the other moved across the surface. This is the linking of two full cells, one kept constant (the fixed cell and the steel directly below it), while the other half cell moves, changing the steel to concrete half cell and thus the second full cell. We will not know the absolute value to the steel to concrete half cell, but we will measure how it changes from point to point. Interpretation of data from fixed versus moving half cell surveys is more difficult than interpreting absolute values, but again, the contour plot is probably the most useful method of interpretation.

4.8 CARBONATION DEPTH MEASUREMENT

Carbonation depth is easily measured by exposing fresh concrete and spraying it with phenolphthalein indicator solution. The carbonation depth must then be related to the cover (the average and its variation) so that the extent to which carbonation has reached the rebar can be estimated, and the future carbonation rate estimated.

There was some discussion in Section 3.1 about whether the carbonation front is truly as well defined as the indicator shows it to be; but for most practical considerations it is a very accurate and reliable technique. Some dark coloured fine aggregates can cause problems by making the colour transition difficult to see. Very poorly consolidated concrete and concrete underground exposed to dissolved carbonates in the water may not show clearly defined carbonation fronts due to the non-uniform progress of the carbonation front.

The cutting up of cores to expose a fresh surface should be done carefully to prevent dust from carbonated areas contaminating the uncarbonated surface and vice versa.

4.8.1 Equipment and use

Carbonation is easily measured by exposing fresh concrete and spraying on phenolphthalein indicator as shown in Figure 4.8. This can be done either by breaking away a fresh surface (e.g. between the cluster of drill holes used for chloride drilling as described in Section 4.9), or by coring and splitting or cutting the core in the laboratory. The phenolphthalein solution will remain clear where concrete is carbonated and turn pink where concrete is still alkaline.

The best indicator solution for maximum contrast of the pink colouration is a solution of phenolphthalein in alcohol and water, usually 1 g indicator in 100 ml of alcohol/water (50:50 mix) or more alcohol to water (Building Research Establishment, 1981; Parrott, 1987). If the concrete is very dry then a light misting with water before applying the phenolphthalein will also help show the colour. Care must be taken that dust from drilling, coring or cutting does not get on the treated surface. Other indicators such as thymolphthalein, Alizarin yellow and universal indicator have been used, along with pH meters. However, phenolphthalein is the most reliable, convenient and widely used indicator (Parrott, 1987).

4.8.2 Interpretation

Carbonation depth sampling can allow the average and standard deviation of the carbonation depth to be calculated. If this is compared

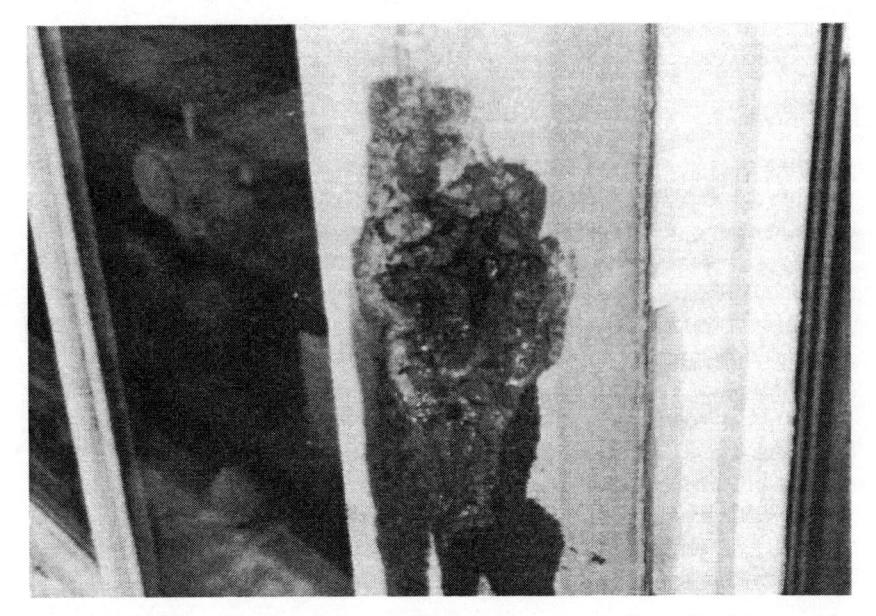

Figure 4.8 Carbonation test on a reinforced concrete mullion. In this case the carbonation had only penetrated about 5 mm on an element with a 25 mm cover. Phenolphthalein solution sprayed on to the freshly exposed concrete shows pink in the alkaline areas (the darker central area around the rebar), and remains clear in the carbonated areas near the surface.

with the average reinforcement cover then the amount of depassivated steel can be estimated. If the carbonation rate can be determined from historical data and laboratory testing then the progression of depassivation with time can be calculated.

4.8.3 Limitations

Phenolphthalein changes colour at pH 9. The passive layer breaks down at pH 10–11. If the carbonation front is 5 to 10 mm wide, the steel can be depassivated 5 mm away from the colour change of the indicator as shown in Figure 3.1. This should be considered when using phenolphthalein measured carbonation depths to determine rate and extent of depassivation.

Some aggregates can confuse phenolphthalein readings. Some concrete mixes are dark in colour and seeing the colour change can be difficult. Care must be taken that no contamination of the surface occurs from dust and the phenolphthalein sprayed surface must be freshly exposed or it may be carbonated before testing.

It is also possible for the phenolphthalein to bleach at very high pH, e.g. after chloride removal (or possibly realkalization). If the sample is left for a few hours it will turn pink. There can also be problems on below ground structures where carbonation by ground water does not always produce the clear carbonation front induced by atmospheric CO_2 ingress.

The number of test areas is usually limited by the amount of concrete that those responsible for the structure will allow to be broken off or holes drilled and cores taken. It is often necessary to do such sampling away from public gaze so it is rarely possible to carry out carbonation tests systematically. They are usually done at accessible locations where damage is not too obvious and repair is easy.

4.9 CHLORIDE DETERMINATION

Chlorides are usually measured by dissolving powder samples in acid. The samples are taken from drillings or from crushed cores. It is preferable to collect a series of drillings at different depths so that a chloride profile can be produced. Alternatively a core can be cut into slices and the slices crushed. Like the carbonation tests the chloride profile must be related to cover so that the extent to which rebars are exposed to high chlorides can be determined.

Chloride profiles can also be used to determine the diffusion coefficient and thus predict the ongoing rate of ingress.

The corrosion thresholds were discussed in Section 3.2.3. It is important to recognize that these are approximate. It is also important to realize that the chloride level at the rebar determines the present extent of corrosion, but the profile (Figure 3.2) determines the future rate, as that is what drives more chlorides from the concrete surface into the steel surface.

Chloride contents can be measured by several methods. In the laboratory, powdered samples are usually digested in acid and then titrated to find the concentration in the conventional wet chemical method.

In the field there are two well known methods of measuring chlorides: Quantab strips and specific ion electrodes. The former are of modest accuracy and can be made inaccurate by certain aggregate types. The latter can be highly accurate. However, the equipment is expensive and requires training and a good methodology to use effectively. Any field technique should be checked against laboratory analysis of duplicate samples.

The results of the above methods are referred to as the 'total' or 'acid soluble' chloride contents. There are also methods for measuring the 'free chlorides' or water soluble rather than the acid soluble chlorides. This refers to the fact that it is the chloride dissolved in the pore water that contributes to the corrosion process. Any chlorides chemically bound up in the cement paste (chloroaluminates or C_3A), or bound up in the aggregates are 'background' chlorides that should not contribute to the corrosion threshold (Section 3.5).

Unfortunately the water soluble techniques produce results that are difficult to reproduce so they are rarely used in the UK or Europe. However, the 'Soxhlet extraction technique', a method of refluxing concrete chips in boiling water to extract the chloride is a standard technique used in North America. For example, AASHTO T260 *Sampling and Testing for Total Chloride Ion in Concrete and Concrete Raw Materials* includes a procedure for water soluble chloride ion analysis, covering sample preparation and analysis.

Chlorides can be cast into concrete or can be transported in from the environment. The chloride ion attacks the passive layer even though there is no significant, generalized, drop in pH (pits in the steel can become very acidic, see Section 2.3). Chlorides act as catalysts to corrosion. They are not consumed in the process but help to break down the passive layer of oxide on the steel and allow the corrosion process to proceed quickly.

Chloride testing will show:

1. Whether chlorides are present in a high enough concentration to cause corrosion. Typically concrete with more than 0.4% chloride by weight of cement is at risk of corrosion.
2. Whether chlorides were cast in or diffused in later. The spatial distribution of chloride with depth and about the structure will show a profile with depth if chlorides diffused in. Its distribution across the surface may be related to areas of rundown (bridge substructures) or splash (marine piles, bridge columns beside roads that are salted in winter). If there is either even distribution, random distribution or one that looks more closely related to the (cold weather) construction schedule than any exposure to external sources of chlorides then chlorides could have been cast in either as a rapid setting agent or been introduced during construction in contaminated water or aggregates.

4.9.1 Property to be measured

The amount of chloride ion in the concrete can be measured by sampling the concrete and carrying out chemical analysis (titration) on a liquid extracted from the sample. The analysis is usually done by mixing acid with drillings or crushed core samples. An alternative is pore extraction by squeezing samples of concrete or, more usually mortar. This technique is frequently used in laboratory experimental work as it is often difficult to extract useful pore water samples from field concrete. The Soxhlet extraction technique for free chlorides was discussed above.

Considerable work has gone into differentiating between bound and free chlorides. As only the free chlorides contribute to corrosion these are ideally what we want to know about. However, the binding of chlorides is a reversible and dynamic reaction, so attempts to remove and measure free chlorides will release bound chlorides. A further complication is that carbonation breaks down chloroaluminates, thus freeing chlorides which proceed as a wave ahead of the carbonation front.

The most accurate and reproducible tests are the acid soluble chloride tests that effectively measure total chlorides. Pore water extraction and water soluble chloride measurements are less reproducible and less accurate. However, from a practical point of view the chloride threshold values are based on total chloride levels so the system is at least self-consistent.

4.9.2 Equipment and use

The collection of chloride samples should be done incrementally from the surface either by taking drillings or sections from cores. The first 5 mm is usually discarded for being directly influenced by the immediate environment. This first increment can show excessively high levels if salt has just deposited on the surface or excessively low levels if rain or other water has just washed away the chlorides. Care should be taken to minimize cross-contamination of samples at different increments.

Measurements of chloride content are made at suitable increments, typically 2 to 5 mm. For improved statistical accuracy when taking drillings, multiple adjacent drillings are made and the depth increments from each drilling are mixed. Special grinding kits are available and sample sizes required for analysis vary from 10 to 50 g (Figure 4.9).

The major concern with sample size is ensuring that there is a uniform amount of cement paste in each sample and that there is no risk of the sample being dominated by a large piece of aggregate. Some researchers have crushed, cored and removed the larger aggregate pieces, measuring only the paste and small aggregates. This is time consuming and the sample is no longer a representative sample as the removal of aggregates cannot be done quantitatively.

There are several ways of measuring the chlorides once samples are taken. Field measurements of acid soluble chloride can be made using a chloride specific ion electrode (Herald et al., 1992). Conventional titration by BS1881, Part 124 and potentiometric titration methods are also available (Grantham, 1993).

Figure 4.9 The RCT profile grinding kit used in the lab to evaluate the chloride diffusion coefficient of a concrete mix. The grinding may also be done *in situ*.

As well as acid soluble chlorides there are the water soluble chloride tests (ASTM D1411, 1982; AASHTO T260, 1984). These techniques use different levels of pulverization of large samples that are refluxed to extract the supposedly unbound chlorides. These are the chlorides that are free in the pore water to cause corrosion as opposed to the chloride bound by the aluminates in the concrete. Further complications arise because some aggregates of marine origin contain chemically bound chlorides within the aggregates even after washing. These are permanently bound, but will show up in the total chloride analysis if the aggregate is attacked by the acid digestion process, or if the aggregate is broken up in the grinding down process. The water soluble chloride test is rather inaccurate as the bound chlorides can be released and the finer the grinding the more will be extracted. However, this test can be useful in showing the corrosion condition where chlorides have been cast into concrete, and particularly where aggregates are known to contain chlorides that do not leach out into the pore water.

4.9.3 Interpretation

There is a well known 'chloride threshold' for corrosion given in terms of the chloride/hydroxyl ratio (Hausmann, 1967). When the chloride concentration exceeds 0.6 of the hydroxyl concentration the passive layer will break down. This approximates to a concentration of 0.2 to 0.4% chloride by weight of cement, 1 lb yd^{-3} of concrete or 0.05% chloride by weight of concrete. The threshold is discussed in Section 3.2.3.

The important questions from chloride measurement are how much of the rebar is depassivated and how will this progress. Points (1) to (3) in Section 3.2.3 review how the corrosivity of the chloride can change. If chlorides have been transported in from outside then the chloride profile can be used along with measurements (or estimates) of the diffusion constant to estimate future penetration rates and the build up of chloride at rebar depth.

Methods of predicting chloride diffusion and the movement of the chloride threshold are discussed in Section 8.2.

4.10 RESISTIVITY MEASUREMENT

Since corrosion is an electrochemical phenomenon, the electrical resistivity of the concrete will have a bearing on the corrosion rate of the concrete as an ionic current (electric current in the form of a flow of ions) must pass from the anodes to the cathodes for corrosion to occur.

The four-probe resistivity meter or Wenner probe was developed for measuring soil resistivity. Specialized modifications of the Wenner probe are frequently used for measurement of concrete resistivity on site. The measurement can be used to indicate the possible corrosion activity if steel is depassivated. Proprietary versions of the system are shown in Figures 4.10(a) and 4.10(b) and a schematic given in Figure 4.10(d). Current is applied between the two outer probes and the potential difference measured across the two inner probes. This approach eliminates any effects due to surface contact resistances.

For a semi-infinite, homogeneous material the resistivity ρ is given by:

$$\rho = 2\pi a V / I$$

where a is the electrode spacing, I is the applied current across

58   *Condition evaluation*

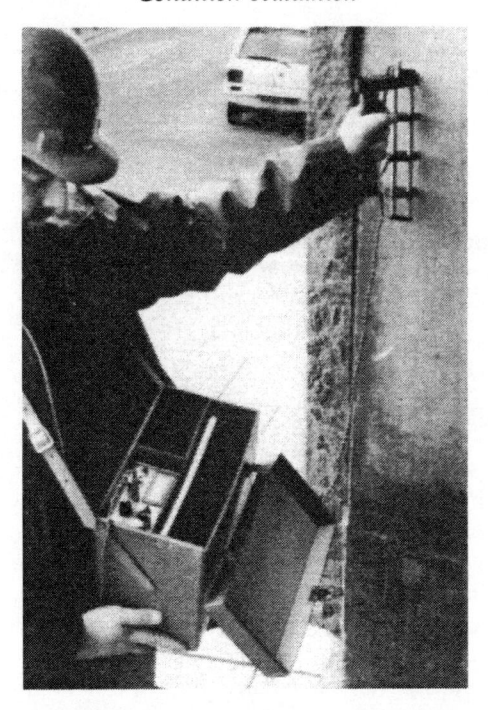

Figure 4.10(a) A four-probe resistivity meter. Courtesy of CNS Electronics.

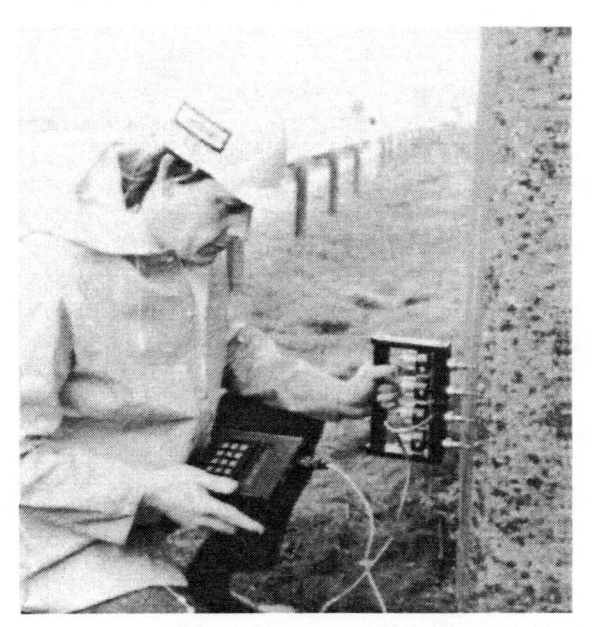

Figure 4.10(b) Four-probe logging resistivity meter with gel contact. Courtesy of Colebrand Ltd.

Figure 4.10(c) A two-probe resistance meter. Courtesy of CMT (Instruments) Ltd.

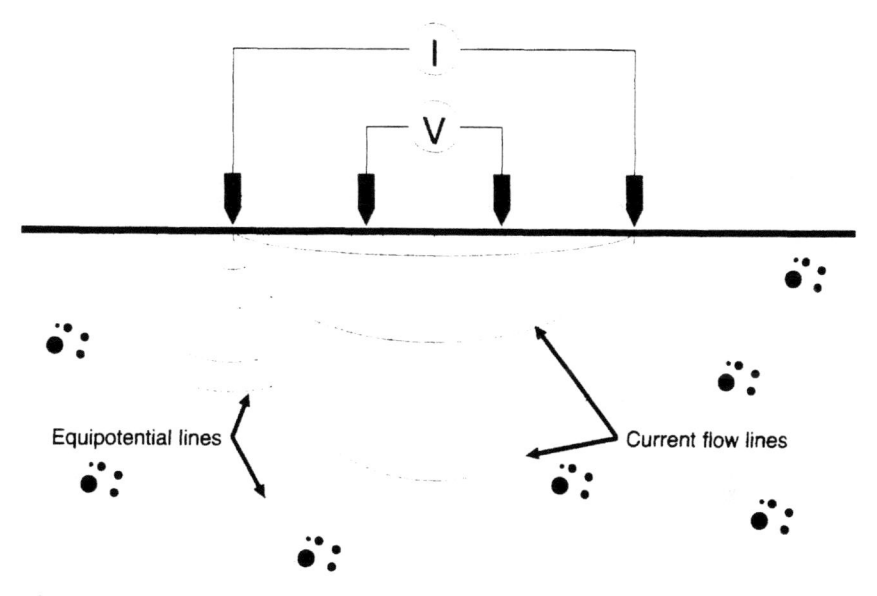

Figure 4.10(d) The four-probe Wenner-type resistivity measurement.

the outer probes and V is the potential measured across the inner probes.

Cheaper, less accurate two probe systems are also available (Figure 4.10(c)). These are often inserted into drilled holes in the concrete to improve electrical contact by getting below the surface latence and any minor carbonation. At one time it was considered necessary to drill

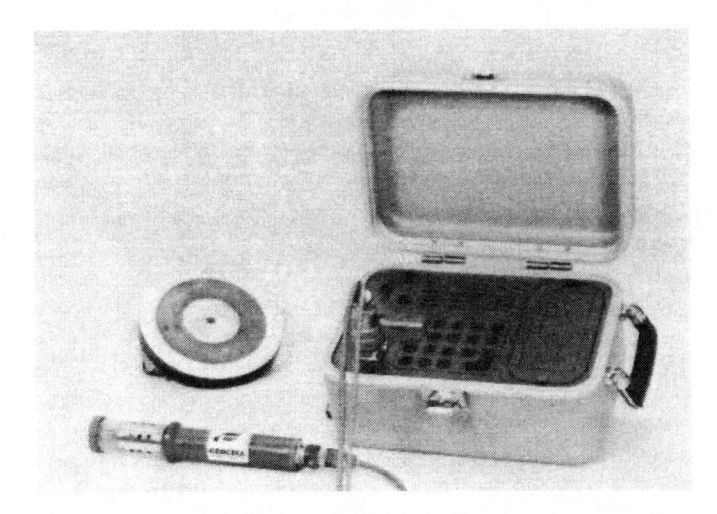

Figure 4.11 The Gecor 6 linear polarization device with sensor controlled guard ring for corrosion rate measurement and concrete cover resistance/resistivity meter. Courtesy Geocisa SA, Spain.

holes to insert the probes, but modern four-probe devices are spring loaded and just push on to the concrete surface. In one commercial device a wetting gel is automatically applied as the probes are pushed on to the concrete surface; in another commercial model wooden plugs in the end of the probes are wetted.

A newer approach using a single electrode on the surface and the rebar network can be used to measure the resistivity of the concrete cover. This is available as part of a corrosion rate measuring device (see Section 4.11 and Figure 4.11) and uses the reinforcement cage as one electrode and a small surface probe as the other electrode. The advantage of this approach is that it measures the resistivity of the cover concrete only. The disadvantage is that it suffers from contact resistance problems.

4.10.1 Property to be measured

The electrical resistivity is an indication of the amount of moisture in the pores, and the size and tortuosity of the pore system. Resistivity is strongly affected by concrete quality, i.e. cement content, water–cement ratio, curing and additives used. The chloride level does not affect resistivity directly as there are plenty of ions dissolved in the pore water already and a few more chloride ions here or there do not make a difference. However, chlorides in concrete can be hygroscopic, i.e. they

Resistivity measurement

will encourage the concrete to retain water. This is why chlorides are often accused of reducing concrete resistivity.

4.10.2 Equipment and use

Millard (1991) has described two commercially available versions of the equipment. Some variations use drilled in probes or a simpler, less accurate two-probe system. It is generally agreed that the four- (or two-) probe system needs a probe spacing larger than the maximum aggregate size to avoid measuring the resistivity of a piece of aggregate rather than of the paste and aggregate. If it is not possible to avoid the influence of reinforcing steel then readings should be taken at right angles to the steel rather than along the length of it as it can provide a 'short circuit' path for the current during the measurement (Figure 4.12). Measurements should also be taken away from edges of the concrete.

The alternative approach measures the resistivity of the cover concrete by a two electrode method using the reinforcing network as one electrode and a surface probe as the other (Broomfield *et al.*, 1993, 1994; Newman, 1966; Feliú *et al.*, 1988).

Best position, minimal interaction with rebars

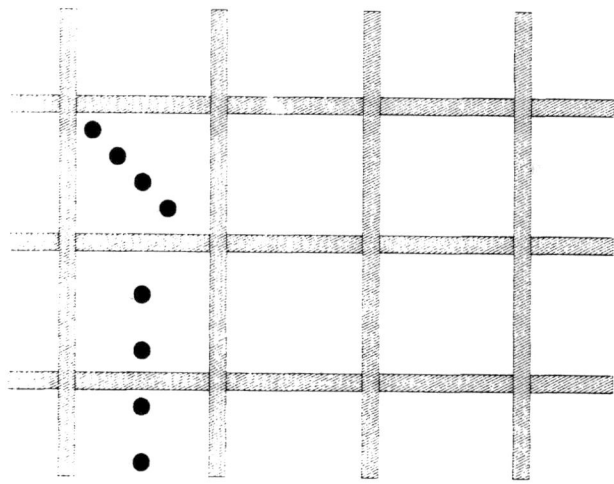

Least interaction if probe array is greater than bar spacing

Figure 4.12 Minimizing the effect of the steel on four-probe resistivity measurements.

Concrete resistivity of the area around the sensor is obtained by the formula:

$$\text{Resistivity} = 2RD \ (\Omega \text{ cm})$$

where R is the resistance by the 'iR drop' from a pulse between the sensor electrode and the rebar network (see Section 6.5 for a discussion of the iR drop) and D is the electrode diameter of the sensor in centimetres. This approach requires a damp surface and the probe is best situated between bars rather than directly over them.

4.10.3 Interpretation

Interpretation is empirical. The following interpretation of resistivity measurements from the Wenner four-probe system has been cited when referring to depassivated steel (Langford and Broomfield, 1987):

> 20 kΩ cm Low corrosion rate
10–20 kΩ cm Low to moderate corrosion rate
5–10 kΩ cm High corrosion rate
< 5 kΩ cm Very high corrosion rate

Researchers working with a field linear polarization device for corrosion rate measurement have conducted laboratory and field research and found the following correlation between resistivity and corrosion rates using the two electrode surface to rebar approach (Broomfield et al., 1993):

> 100 kΩ cm Cannot distinguish between active and passive steel
50–100 kΩ cm Low corrosion rate
10–50 kΩ cm Moderate to high corrosion where steel is active
< 10kΩ cm Resistivity is not the controlling parameter

In the above method and interpretation the resistivity measurement has been most widely used as part of a corrosion rate measurement system and is used alongside linear polarization measurements (see Section 4.11), not as a stand alone technique.

4.10.4 Limitations

Resistivity measurement is a useful additional measurement to aid in identifying problem areas or confirming concerns about poor quality concrete. Readings can only be considered alongside other measurements.

There is a temptation to multiply the resistivity by the half cell potential and present this as the corrosion current or rate. This is incorrect. The corrosion rate is usually determined by the interfacial resistance between the steel and the concrete, not the bulk concrete resistivity. The potential measured by a half cell is not the potential at the steel surface that drives the corrosion cell. Correlations between resistivity, half cell potential and corrosion rate may be found in similar samples in similar conditions in the laboratory but in the variability of the real world any correlation is fortuitous.

The main problem with the four-probe technique is that the reinforcing steel will provide a 'short circuit' path and give a misleading reading. However, research at the University of Liverpool has shown that if measurements are taken at right angles to a single reinforcing bar the error is minimized as shown in Figure 4.12.

Obviously corrosion is an electrochemical process with current in the form of ions flowing through the concrete. The resistivity can tell us the capacity of the concrete to allow corrosion. It will not tell us if corrosion has started or if that capacity is being used to the full. Hence the statement above that at less than 10 kΩ cm resistivity is not the controlling parameter, but at more than 100 kΩ cm you cannot distinguish between active and passive steel as the resistivity will effectively stop corrosion.

The resistivity calculation assumes the concrete to be homogeneous. The local inhomogeneity of the aggregate must be allowed for by suitable probe spacing. The systematic inhomogeneity of the reinforcement network must be allowed for by minimizing its effect as shown in Figure 4.12. There is another effect due to layers of different resistivity caused by carbonation, water ingress, etc. This is discussed in Millard (1991) and an 'influence diagram' given to show the effect of layers of different resistance on the measurement.

4.11 CORROSION RATE MEASUREMENT

The corrosion rate is probably the nearest the engineer will get with currently available technology to measuring the rate of deterioration. There are various ways of measuring the rate of corrosion, including AC impedance and electrochemical noise (Dawson, 1983). However, these techniques are not suitable for use in the field for application to the corrosion of steel in concrete so this section will concentrate on linear polarization, also known as polarization resistance, and will discuss various macrocell techniques.

Unfortunately we cannot yet use corrosion rate measurements to calculate total steel section loss or predict concrete spalling rates. However, we can tell the engineer how much steel is turning into rust and how much metal is being lost at the time of measurement. The use of a corrosion model to predict deterioration rates using chloride profiles, carbonation depths and corrosion rate measurements is discussed in Chapter 8.

4.11.1 Property to be measured

It is possible, with varying degrees of accuracy, to measure the amount of steel dissolving and forming oxide (rust). This is done directly as a measurement of the electric current generated by the anodic reaction:

$$Fe \rightarrow Fe^{2+} + 2e^-$$

and consumed by the cathodic reaction:

$$H_2O + \tfrac{1}{2}O_2 + 2e^- \rightarrow 2OH^-$$

and then converting the current flow by Faraday's law to metal loss:

$$m = Mit/zF$$

where m is the mass of steel consumed, i is the current (amperes), t is the time (s), F is 96 500 A s, z is the ionic charge (2 for $Fe \rightarrow Fe^{2+} + 2e^-$) and M is the atomic weight of metal (56 g for Fe). This gives a conversion of 1 $\mu A\ cm^{-2}$ = 11.6 μm steel section loss per year.

4.11.2 Equipment and use: linear polarization

The linear polarization technique requires us to polarize the steel with an electric current and monitor its effect on the half cell potential. It is carried out with a sophisticated development of the half cell incorporating an auxiliary electrode and a variable low voltage DC power supply. The half cell potential is measured and then a small current is passed from the auxiliary electrode to the reinforcement. The change in the half cell potential is simply related to the corrosion current by the equation:

$$I_{corr} = B/R_p \qquad (4.1)$$

where B is a constant (in concrete 26 to 52 mV depending upon the passivity or active condition of the steel) and R_p is the polarization resistance (in ohms):

$$R_p = \text{(change in potential)}/\text{(applied current)} \qquad (4.2)$$

R_p gives the technique its alternative name of 'polarization resistance'. The change in potential must be kept to less than 20 mV or so for the equation to be valid and remain linear (hence the name 'linear polarization'). The 'iR drop' must also be removed. This is the voltage that exists because a current is flowing through concrete that has an electrical resistance. This is also referred to as the 'solution resistance'. This means that the current is usually switched off during the measurement process so that the potential without the iR drop is measured. The iR drop is discussed further in Section 6.5 on cathodic protection criteria.

The measurement is made in one of two ways. Either steady fixed levels of current are applied and the potential monitored (galvanostatic), or the current is increased to achieve one or more target potentials (potentiostatic). In both cases allowances for the iR drop (solution or concrete resistance) must be made. A plot of change in current versus change in potential gives a gradient of the polarization resistance R_p (equation 4.2) to calculate the steel section loss rate.

The corrosion rate in $\mu m \ yr^{-1}$ is:

$$x = (11 \times 10^6 \ B) / (R_p \ A) \quad (4.3)$$

where A is the surface area of steel measured in square centimetres.

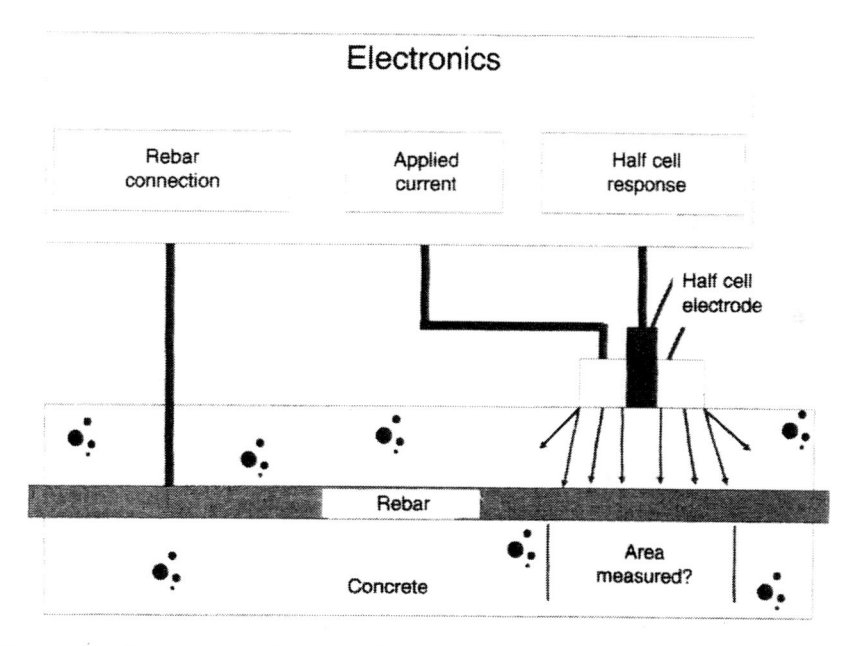

Figure 4.13 Schematic of the set-up for linear polarization.

A schematic of a typical linear polarization device is shown in Figure 4.13.

Defining the area of measurement is crucial to accurate corrosion rate measurement. In other applications linear polarization is carried out on a sample of known size in a pipe, tank or condenser where corrosion is a risk. However, we cannot put a sample into concrete, we must use the steel that is there which is all connected together.

The area of steel sensed by the electrodes is obviously far smaller than the whole rebar network. However, we cannot assume that the area measured is that directly below the electrode as the current 'fans out' as shown in Figure 4.13. Assuming a 1:1 relationship between probe size and area of steel polarized can lead to errors of up to 100× in our measurement (Fliz et al., 1992).

In one commercial device a guard ring system has been developed to confine the area of the impressed current and thus define the parameter A in equation 4.3 which allows us to calculate the metal loss. This is illustrated schematically in Figure 4.14; the equipment is shown in Figure 4.11. Less accurate devices use a large electrode and assume that area of measurement is directly below the electrode. Some devices have guard rings that apply fixed levels of current which do not always confine the current correctly. These approaches are reasonably accurate

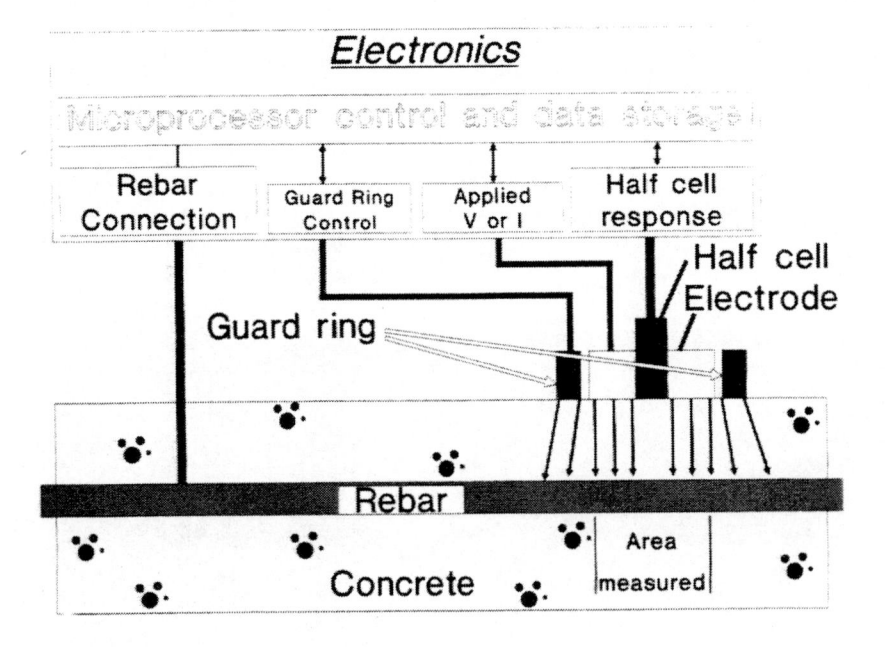

Figure 4.14 Schematic of linear polarization device with sensor controlled guard ring for defining area measured.

at high corrosion rates but less accurate when corrosion rates are low or localized (which is when you may require greatest accuracy).

Criteria relating linear polarization measurements to deterioration rates, similar to the ASTM C876 (1991) criteria for half cell potentials, have been published (Broomfield et al., 1993). These show some comparability between different devices and will be discussed below under 'interpretation' (Section 4.11.4). A set of conversion equations is provided in the final report of the Strategic Highway Research Program (SHRP) contract on corrosion rate measurement (Fliz et al., 1992).

4.11.3 Carrying out a corrosion rate survey

Corrosion rate measurement is slow compared with half cell potential measurement. This is because the concrete reacts slowly to the electric field and changes must be reasonably slow to ensure that the electrochemistry in the concrete is changing linearly and without capacitance effects. However, the speed of the total operation varies significantly from device to device. The slowest devices are manually operated and take 10–20 minutes to take a reading that then must be manually plotted to calculate the corrosion current. The fastest microprocessor controlled devices take less than five minutes and give the corrosion current directly.

As the technique is slower than taking half cell measurements it is important to take measurements at the most significant locations on the structure, e.g. by following up a potential survey with strategic corrosion rate measurements. Rate measurements should be taken at the positions of the highest and lowest potentials and at the steepest potential gradients.

It should also be noted that corrosion rates vary with the weather conditions. Corrosion rates increase in warm conditions. Resistivity will decrease as the concrete gets wet, also allowing corrosion rates to increase. For a full picture of corrosion conditions, measurements should be taken at regular intervals throughout the year so that seasonal changes can be identified. Alternatively readings can be taken at the same time each year, preferably under comparable weather conditions, so that results are comparable.

The corrosion rate is measured over a known area of rebar. This means that the rebars must be located and their sizes known so that the area of steel below the sensor is known. For purely comparative readings it can be adequate to take readings in comparable locations (such as isolated sections of bars or a cross-over).

The potential must be stable throughout the reading so that a true change in potential (equation 4.2) is recorded. This can lead to problems on very dry structures and where conditions are changing rapidly. Local wetting and coming back to the problem location later can sometimes overcome these problems.

Gowers et al. (1992) have used the linear polarization technique with embedded probes (half cell and a simple counter electrode) to monitor the corrosion of marine concrete structures. This technique was described previously without reference to isolating the section of bar to be measured (Langford and Broomfield, 1987). By repeating the measurement in the same location on an isolated section of steel of known surface area the corrosion rate of the actual rebar can be inferred. The main problem is the long-term durability of electrical connections in marine conditions. Also the probes should ideally be built into the structure during construction – a desirable but rare occurrence.

4.11.4 Interpretation: linear polarization

The following broad criteria for corrosion have been developed from field and laboratory investigations with the sensor controlled guard ring device shown in Figures 4.11 and 4.14 (Broomfield et al., 1993, 1994):

$$\text{Passive condition: } I_{corr} < 0.1 \text{ } \mu\text{A cm}^{-2}$$
$$\text{Low to moderate corrosion: } I_{corr} \text{ } 0.1 \text{ to } 0.5 \text{ } \mu\text{A cm}^{-2}$$
$$\text{Moderate to high corrosion: } I_{corr} \text{ } 0.5 \text{ to } 1 \text{ } \mu\text{A cm}^{-2}$$
$$\text{High corrosion rate: } I_{corr} > 1 \text{ } \mu\text{A cm}^{-2}$$

The device without sensor control has the following recommended interpretation (Clear, 1989):

$$\text{No corrosion expected: } I_{corr} < 0.2 \text{ } \mu\text{A cm}^{-2}$$
$$\text{Corrosion possible in 10–15 years: } I_{corr} \text{ } 0.2 \text{ to } 1.0 \text{ } \mu\text{A cm}^{-2}$$
$$\text{Corrosion expected in 2–10 years: } I_{corr} \text{ } 1.0 \text{ to } 10 \text{ } \mu\text{A cm}^{-2}$$
$$\text{Corrosion expected in 2 years or less: } I_{corr} > 10 \text{ } \mu\text{A cm}^{-2}$$

These measurements are affected by temperature and relative humidity, so the conditions of measurement will affect the interpretation of the limits defined above. The measurements should be considered accurate to within a factor of two.

It should be noted that different values of B are used by the two devices (equation 4.1), with 26 mV used by the guard ring device and 52 mV used by the device without a guard ring. This may explain the factor of two difference in interpretation at the low end. At the high end, the lack of a guard ring may lead the simpler device to include a larger area of steel in its measurement, indicating a higher corrosion rate. Alternatively the device may have been used on more actively corroding structures and the interpretation range may therefore have been extended. An equation linking the devices and a third, simple guard ring, is given in Fliz et al. (1992). The correlation factors (R^2) showed relative accuracies of the devices of 70–90%.

Effects of temperature

Temperature affects the corrosion rate directly. The rate of the oxidation reaction is affected by the amount of heat energy available to drive the reaction. The concrete resistivity also reduces with increased temperature as ions become more mobile and salts become more soluble. It affects the relative humidity in the concrete, lowering it when the temperature increases and therefore introducing an opposite effect. At the low end of the temperature scale the pore water will freeze and corrosion stops as the ions can no longer move. It should be noted that this will happen well below ambient freezing point as the ions in the pore water depress the freezing point well below 0°C.

Effects of relative humidity (RH)

This will also be a factor in determining how much water there is in the pores to enable the corrosion reaction to be sustained. Chloride induced corrosion is believed to be at a maximum when the RH within the concrete is around 90–95% (Tuutti, 1982). For carbonation there is experimental evidence that the peak is around 95–100% RH. However, it is important to recognize that RH in the pores is not simply related to atmospheric RH; water splash, run off or capillary action, formation of dew, solar heat gain or other factors may intervene.

Increased water saturation will slow corrosion through oxygen starvation because once the pores are filled with water, oxygen cannot get in. Conversely, totally dry concrete cannot corrode. However, when concrete is close to saturation the embedded steel can corrode rapidly without signs of cracking. This is due to the limited amount of oxygen available as the iron ions (Fe^{2+}) stay in solution without forming solid rust that expands and cracks the concrete. This is probably the condition that lead to the severe loss of section on the bars shown in Figure 2.3. A totally saturated structure can reach a very high negative half cell potential (–900 mV) without corroding, as a result of oxygen starvation. However, if oxygen does then get access to the steel the corrosion rate will be very high as the steel will have no passive oxide layer to protect it. This was discussed in Section 4.7.2.

Corrosion rate measurement can be very important in finding the true situation in such ambiguous conditions.

Pitting

Much of the research on linear polarization has been done on highway bridge decks in the USA where chloride levels are high and fairly uniform across the structure. Because of the uniform distribution of chloride and the open exposed surface with good oxygen access pitting is not observed (see Section 2.3). However, in Europe, corrosion may be localized in run down areas on bridge substructures. There is ponding of water on the beam surface, and salt water can get under membranes due to damage or the end of the membrane. These circumstances can lead to pitting. The problems of interpretation of corrosion rate readings are complicated as the I_{corr} reading comes from isolated pits rather than uniformly from the area of measurement.

Laboratory tests with the guard ring device have shown that the corrosion rate in pits can be up to ten times higher than generalized corrosion. This means that the device is very sensitive to pits. However, linear polarization devices cannot differentiate between pitting and generalized corrosion. That must be done by direct observation of the steel or by careful study of the half cell potentials (Vassie, 1991) and chloride contents.

Carbonation versus chlorides

Most work on linear polarization probes has been done in chloride corrosion condition. However, the only methods of assessing carbonated concrete are destructive drilling or coring for carbonation depth measurement and trying to interpret half cell potentials which is difficult (Section 4.7.2). Linear polarization is therefore very useful in assessing carbonated structures, particularly as half cell potentials are so difficult to interpret for carbonation induced corrosion.

The corrosion rates associated with carbonation rarely exceed $0.5\,\mu A\,cm^{-2}$ while chloride attack can give corrosion rates of more than $1\,\mu A\,cm^{-2}$ equivalent to $12.5\,\mu m$ section loss per year metal (Broomfield et al., 1993, 1994 and Figure 4.15). This may be due to the fact that carbonated concrete dries out more rapidly than chloride contaminated concrete and surveys are normally done in the dry, when the rate would be lower in the carbonated condition.

Predicting deterioration rates

Measurements can be used to predict the deterioration rate of the structure when used in combination with diffusion data, chloride or carbonation penetration and cover measurements. The ongoing rate of deterioration may be important in deciding whether to defer rehabilita-

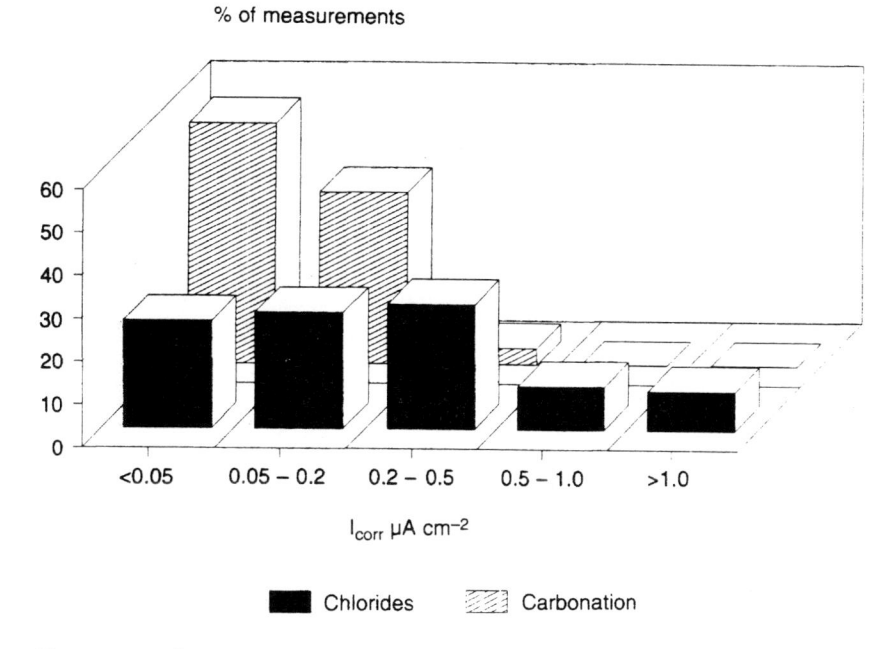

Figure 4.15 Comparison of corrosion rates (current) due to carbonation and chloride attack (I_{corr} values).

tion work, carry out minor cosmetic repairs in anticipation of later major works, or whether to bring forward major works. It may also be used to prioritize repairs either element by element or on a structure by structure basis. There is further discussion of this in Chapter 8.

Some work has been done in translating I_{corr} to section loss and end of service life (see equation 4.3 and Andrade *et al.*, 1990). However, the loss of concrete is the most usual cause for concern, rather than loss of reinforcement strength. It is far more difficult to predict cracking and spalling rates especially from an instantaneous measurement. A simple extrapolation, assuming that the instantaneous corrosion rate on a certain day is the average rate throughout the life of the structure, often gives inaccurate results (Broomfield *et al.*, 1993). This is true for section loss. Converting that to delamination rates is even less accurate as it requires further assumptions about oxide volume and stresses required for cracking.

The best that we can do at present is to work on the estimates of how much section loss will generate enough expansive oxide growth to cause cracking. This is between 10 and 100 μm (0.01 to 0.1 mm).

The oxides produced by iron have volume increases between twice and six times the steel consumed. This assumes that they are 100%

dense (with no porosity occupying extra space). To generate cracking the oxides must be deposited at the metal/oxide interface.

If we assume that the average expansion ratio is 3, allowing for oxide depositing elsewhere and some porosity in the oxide (Figure 3.4), then the corrosion rates given above translate as follows:

$0.1\ \mu A\ cm^{-2} \equiv 1.1\ \mu m\ yr^{-1}$ section loss $\equiv 3\ \mu m\ yr^{-1}$ rust growth

$0.5\ \mu A\ cm^{-2} \equiv 5.7\ \mu m\ yr^{-1}$ section loss $\equiv 17.3\ \mu m\ yr^{-1}$ rust growth

$1.0\ \mu A\ cm^{-2} \equiv 11.5\ \mu m\ yr^{-1}$ section loss $\equiv 34\ \mu m\ yr^{-1}$ rust growth

$10\ \mu A\ cm^{-2} \equiv 115\ \mu m\ yr^{-1}$ section loss $\equiv 345\ \mu m\ yr^{-1}$ rust growth

In laboratory tests (Rodríguez et al., 1994) 15 to 40 μm section loss gave rise to cracking on bars with a cover/diameter ratio between 2 and 4. This is around the 0.5 $\mu A\ cm^{-2}$ transition from low to moderate corrosion rate.

Before and after measurements

Another major application is in defining the effectiveness of a repair. The application of a silane, an inhibitor or of chloride removal or realkalization can lead to the need for long-term monitoring to see if the treatment is still working.

4.11.5 Limitations: linear polarization

There are two major limitations to the linear polarization devices described above:

1. They detect the instantaneous corrosion rate which can change with temperature, relative humidity and other factors.
2. Either they make assumptions about the area of measurement or they define an area of measurement.

The first factor is a limitation because we do not want to know the instantaneous corrosion rate. We actually want to know the integrated corrosion rate from the time of initiation of corrosion. We must therefore make assumptions or multiple measurements with time to estimate the average or integrated rate.

The second limitation is that there can be errors of 10 to 100 in the estimated area of measurement especially at low corrosion rates, unless the sensor controlled system is used. Even if the sensor controlled system is used it gives a single value for the steel contained within a 105 mm circle below the sensor. If corrosion is at a few pits it will underestimate the corrosion rate in the pits and overestimate the general corrosion rate. If it is known that there is pitting corrosion then corrections can be made based on the fact that corrosion rates in pits are five to ten times that of generalized corrosion.

There are further errors because the value of B in equation 4.1 varies from 26 to 52 mV depending upon whether the steel is active or passive. Further, corrosion may be concentrated on the top of the bar or, if bars are close together or deep within the concrete, the device may only send current to the top steel. Both of these errors mean that the best accuracy that can be expected from a linear polarization device is a factor of two to four (Andrade et al., 1995). However, the scale is logarithmic so such errors are less critical than they may seem.

The final limitation is the difficulty of converting the corrosion rate in terms of rate of metal section loss to a total section loss or cracking and spalling rate. Further work in this area is leading to the resolution of this problem.

4.11.6 Equipment and use: macrocell techniques

An alternative approach to the linear polarization technique, which reintroduces the theme of long-term corrosion monitoring, is the embedding of macrocell devices. This includes galvanic couples of different steels (Beeby, 1985) or embedding steel in high chloride concrete to create a corrosion cell, as is popular in cathodic protection monitoring systems, particularly in North America (NACE, 1990).

This approach is also popular in laboratory corrosion studies and has been developed as an ASTM procedure (ASTM G109, 1993). Concrete prisms are made with a single top rebar in chloride containing concrete and two bottom rebars in chloride free concrete. The current flow between the top and bottom is monitored as shown in Figure 4.16.

A recently developed device has been installed in new structures, e.g. the Great Belt Bridge/Tunnel in Denmark (Shiessl and Ruapach, 1993). This uses a 'ladder' of steel specimens of known size, installed at an angle through the concrete cover. As the chloride (or carbonation) front advances each specimen becomes active and the current flow from the anodic specimen to its adjacent cathodic specimen can be monitored along with the half cell potential (Figure 4.17).

4.11.7 Interpretation: macrocell techniques

The macrocell measures the Faraday current from metal dissolution, the same as in the linear polarization technique. There are two major

74   *Condition evaluation*

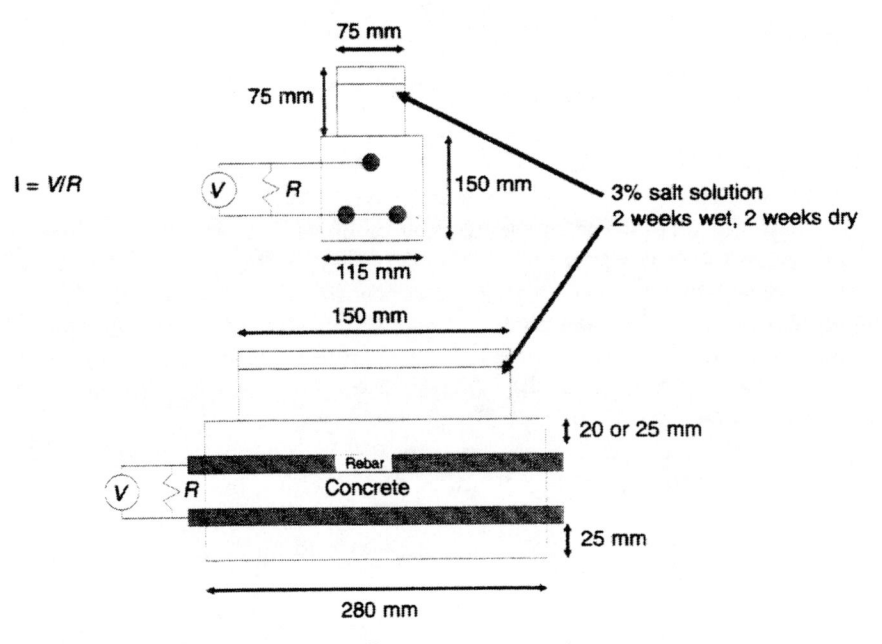

Figure 4.16 ASTM G109 macrocell current prism.

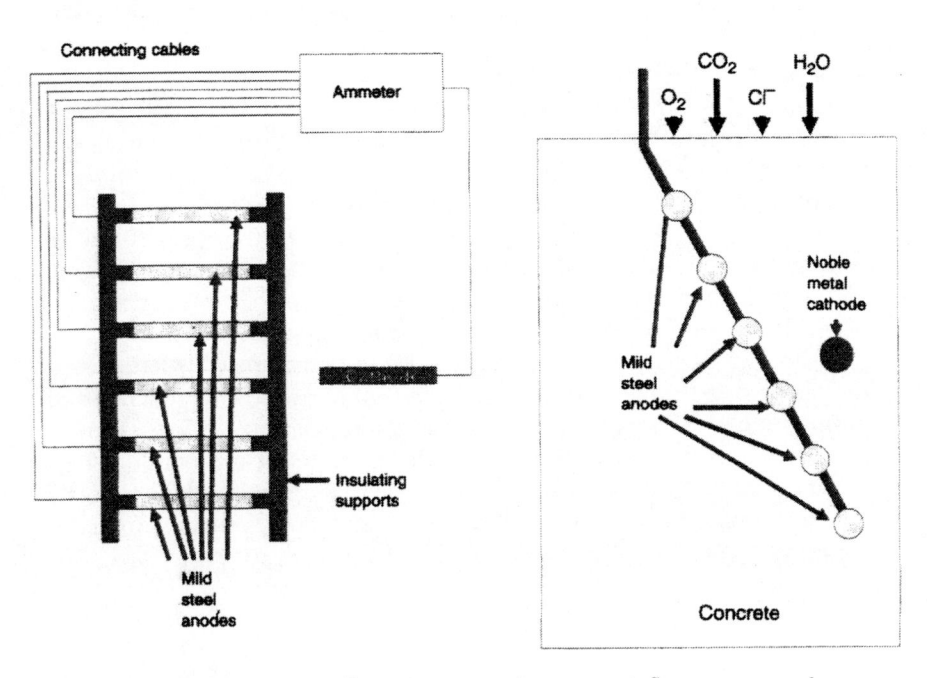

Figure 4.17 Ladder macrocell system corrosion current flow once anodes are depassivated.

problems with interpreting macrocell systems. The first is that we are not actually measuring corrosion on the reinforcing steel of a real structure. We must assume that the current flows in the macrocell probe are representative of what is happening on the steel we want information about. That will partly depend upon the care and thought taken on installation of the probes.

The second issue is how representative macrocell currents are of the true corrosion currents in the steel. The microcell currents may be more important than macrocell current flows. In a comparison with linear polarization (Berke et al., 1990) the macrocell technique underestimated the corrosion rate, sometimes by an order of magnitude. As this was using the ASTM prism technique, it should be considered the most accurate use of the macrocell technique, so if it is an order of magnitude out, field use of macrocell techniques is probably even less accurate.

The advantage of these techniques is that they are permanently set up and can be used to monitor the total charge passing with time. This is a clearer indication of the total metal loss or total oxide produced than an instantaneous measure of the corrosion rate.

4.12 OTHER USEFUL TESTS FOR CORROSION ASSESSMENT

This section briefly summarizes some other tests that are less widely used in corrosion assessment.

4.12.1 Permeability and absorption tests

Since corrosion of steel in concrete is usually caused by the ingress of various agents (Cl^-, CO_2, H_2O, O_2) through the concrete cover, many attempts have been made to calculate and measure the permeability of the concrete. Measurements are most accurately done on a conditioned core in the laboratory, since the environment (e.g. degree of saturation and temperature) will strongly influence field measurements. However, field measurements are made with vacuum devices and water absorption kits to give a rating of the permeability.

Field devices are most effective when used to check improvements after a treatment, for instance the application of a penetrating sealer such as silane. The ISAT test, which has been modified by US researchers, vacuum tests which equate air movement with moisture permeability and other devices all give approximate indications of the rate of diffusion (Whiting et al., 1992; Whiting and Cady, 1993).

In the laboratory, diffusion, permeability, permittivity and absorption can all be measured under controlled conditions. Most are concerned with modelling the flow of chloride into concrete (e.g. ASTM C1202, 1991). Unfortunately even the most accurate measurements in the laboratory can rarely be equated to field conditions where wetting and drying accelerates chloride uptake and the surface concentration is rarely either known or constant, so steady state diffusion is a poor model at best.

'Effective' surface concentrations are used by some US researchers to back-calculate diffusion constants from the measured concentrations. These are usually the concentrations measured at $\frac{1}{4}''$ to $\frac{1}{2}''$ depth (6 to 13 mm), where the chloride concentration is less affected than the surface where recent washing or drying may deplete or enhance the chloride concentration.

An alternative technique is to plot the square root of the effective concentration (total minus background) versus depth, and extrapolating a straight line fit to the Y axis. This method is discussed in Chapters 3 and 8.

4.12.2 Concrete characteristics: cement content, petrography, W/C ratio

It is usually essential to take cores and carry out selective petrographic analysis during the full survey. This can be used to determine the types of aggregates (susceptibility to ASR, freeze–thaw damage, sulphate attack, presence of highly absorptive constituents, etc.), mix design (cement, water and aggregate ratios) and curing and hydration (Concrete Society, 1984).

4.12.3 Radar, radiography, PUNDIT, pulse velocity

Various electromagnetic and acoustic techniques are used for nondestructive examination of concrete. These include radar for locating delaminations, radiography for mapping the rebar network, ultrasonic techniques like pulse velocity and PUNDIT, which can estimate concrete strength and quality and find defects like voids, honeycombing, cracks, etc. All of these techniques require specialized equipment and most require specialists to operate and interpret them. Figures 4.18(a) and (b) illustrate the equipment and use of pulse velocity and impact echo.

The pulse velocity approach is best done in a 'transmission mode' with the pulse created on one side of a concrete member and detected on the other side. It can be used at corners or in a 'reflective' mode if necessary, but it loses effectiveness and interpretation is harder. The impact echo technique can be used with pulse generator and detector side by side as the echo is reflected back from defects.

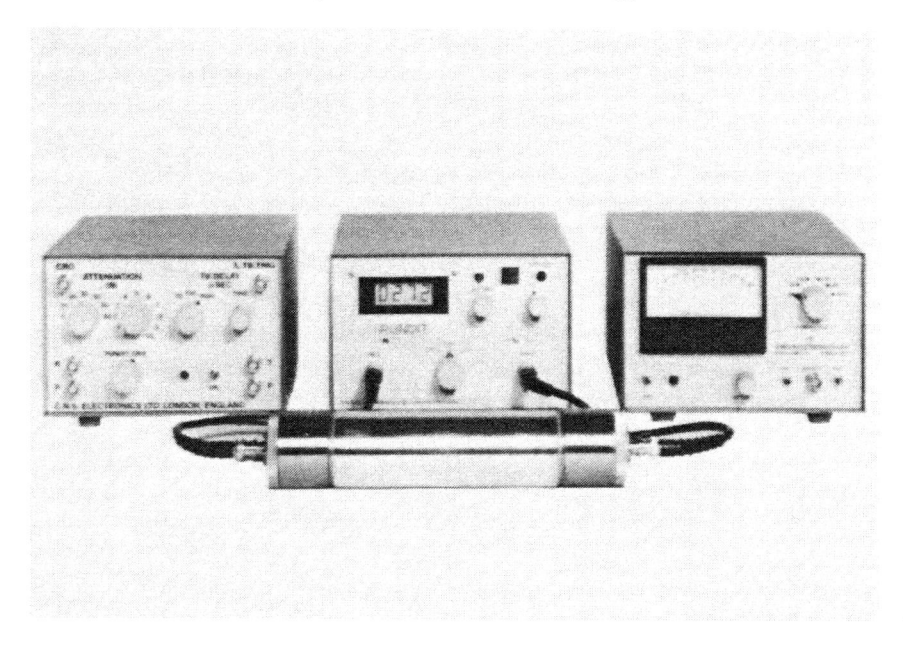

Figure 4.18(a) The PUNDIT pulse velocity device. Courtesy CNS Electronics.

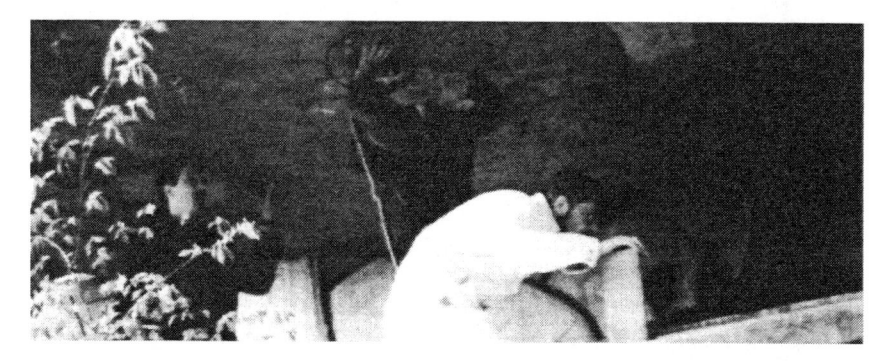

Figure 4.18(b) Testing of a bridge deck with the DOCter impact echo test system. The deck is suffering from ASR reactions, freeze-thaw and starting corrosion causing debonding between the rebars and the concrete.

4.13 SURVEY AND ASSESSMENT METHODOLOGY

When carrying out a survey (or commissioning a specialist test house) it is important to define what information is needed and how that information can be collected accurately but economically. If access is a problem (e.g. on a bridge substructure or a working plant) it may be important to collect available information during a single access period rather than go through the expense or difficulty of providing repeat access.

Plan what measurements will be taken and where they will be taken. Allow for flexibility as the information changes during the survey. Try to check and analyse data from the survey as it proceeds. If potentials are high in one area that may be the best place for coring and drilling (as well as taking some representative samples from other areas).

Plan to spend significant resources on the survey. It could save you from wasting a lot of money later on. Have regard for the repair options available.

If an electrochemical repair technique is under consideration it may be worth checking for the degree of continuity of reinforcement as good continuity is essential but may be difficult or expensive to establish. Time and effort spent at this stage may produce later savings. Conversely, a poorly executed survey can produce problems if a rehabilitation technique based on the survey results turns out to be uneconomic or impractical. Examples that the author has come across include expensive patch repairs and coatings that failed after less than a year leading to later installation of a cathodic protection system, doubling rehabilitation costs, and a cathodic protection system, which had to be redesigned when it was found that there was two to five times as much steel in the structure as originally estimated.

An important aspect, when considering cathodic protection or other electrochemical techniques, is zero cover or tie wire touching the surface. If there is a direct metallic connection between the anode and the rebar it will short out the electrochemical protection. Excessive shorts have caused cathodic protection systems to be abandoned in the USA. On the other hand, the lack of electrical continuity between reinforcing steel that concerns many engineers is rarely a problem and is usually straightforward to overcome. These issues are discussed further in Section 6.6 on cathodic protection.

When planning a survey it is important to know what information you want by the end of it. While executing the survey it is important to interpret the data as it is collected to ensure that the most useful measurements are taken and that new information is used to draw the correct conclusions, not just the expected ones. The survey report should draw conclusions and make recommendations, not just be a summary of the data.

4.14 MONITORING

The technique of corrosion monitoring is well established for cathodic protection of reinforced concrete structures where microprocessor controlled systems are linked by modem to remote monitoring stations and potentials and current flows are monitored (Broomfield et al., 1987). However, very few such systems have been used on actively corroding structures. The sort of system available was described several years ago (Langford and Broomfield, 1987). Variations on this concept were used in the Middle East and on the Midland Links. Recently a system using embedded half cells only was described (Gobbett, 1993). The 'ladder' macrocell system, with variations, is used in new construction where conditions are aggressive and long-term monitoring is required (Shiessl and Ruapach, 1993).

The advantage of long-term monitoring is that the progression of condition changes can be monitored. The growth of anodic areas, using half cells, changes in corrosion rates using linear polarization or macrocell approaches, and the changes in concrete resistivity with time are more helpful in predicting long-term durability than the 'snapshot' approach that a survey entails.

The installation of monitoring systems on new structures is to be recommended, especially where access is difficult, durability is a major issue and for prestigious structures with very long lifetimes where adequate maintenance must be thought out at the design and construction stage (as it should be on all structures).

4.15 SPECIAL CONDITIONS: COATED REBARS AND PRESTRESSING

There are particular problems when the reinforcing steel is not conventional steel bars embedded directly into the concrete. There are problems for galvanized steel bars with respect to half cell and corrosion rate measurement because the zinc affects the readings in poorly understood ways. However, when the bar is coated in epoxy or the reinforcement is in the form of wires in ducts the problems are multiplied, as described below.

4.15.1 Epoxy coated and galvanized reinforcing bars

Epoxy coated rebars present particular problems to determining the corrosion condition of the steel. In the first place the bars are electrically isolated from the concrete except at areas of damage. The sizes and locations of the areas of damage are obviously unknown. Attempts to carry out half cell potential surveys and linear polarization measurements have therefore been unable to come up with definitive criteria for corroding and non-corroding areas. The other problem is that the bars are isolated from each other so a connection must be made to each bar measured to be sure that contact is being made.

When the steel is covered in other coatings, such as the zinc of galvanizing, then the potentials are created by the zinc, not the steel. This leads to very negative potentials (about –500 mV with respect to an

Ag/AgCl or calomel electrode) when the structure is new. This drifts down to smaller values either as the zinc passivates in the alkaline concrete or as it is consumed and the steel becomes active. It can be impossible to distinguish the two conditions, so potential measurements are not useful on galvanized structures.

Corrosion rate measurements should be representative of galvanized steel but the B value in equation 4.1 for zinc may be very different from that of steel in concrete. The author is not aware that any B values for zinc or galvanized steel in concrete have been published. However, comparative results may be useful in showing areas of high and low corrosion. When the zinc is partially lost then the B value will be totally unknown as it will be a mix between that for steel and that of zinc.

4.15.2 Internal prestressing cables in ducts

Pretensioned prestressing systems with wires or rods embedded directly in the concrete can be treated as though they are reinforcing bars, allowing for the different metallurgy. However the problem of prestressing steel cables in internal post-tensioned structures with ducts is more difficult. These structures are usually built of precast elements that are either individually or collectively stressed by cables in ducts running deep within the structures (cover of typically 50 to 100 mm) and then the duct is filled with a cementitious grout.

Investigations around the world have shown extensive problems with poor grouting of the ducts and consequent leakage of water, sometimes containing chlorides, onto the steel cables, with subsequent severe corrosion. Occasionally this has lead to failure of bridges and other structures (e.g. Woodward and Williams, 1988). The problem is that once the steel has been placed 50 to 100 mm or deeper within a structure, surrounded by a steel or polymer conduit, normal non-destructive test techniques are not effective.

Also, as prestressing steel is loaded to more than 50% of its ultimate tensile strength, any section loss, particularly pits, lead to crack initiation and rapid, catastrophic failure. There are apocryphal tales of prestressing rods shooting out of buildings due to corrosion induced failures.

The following test techniques are used in these situations:

1. radiography using a large mobile unit to 'X-ray' the structure and examine for tendon failures;
2. drilling into ducts at two locations and using air flow to check for poor grouting;
3. a new 'magnetic flux' technique that seems able to detect section loss and tendon failure using a high powered magnet and detecting mag-

netic flux leakage (Ghorbanpoor and Shi, 1995); this is still in development;
4. more conventional investigations of the anchorages and ducts to see if they are in good condition or whether any deterioration has occurred that could also have reached the tendons within the ducts.

The failures of prestressed concrete structures and concern about methods of investigating and rehabilitating them are a major issue in the civil engineering and concrete repair industries. The problems of repairing and rehabilitating these structures are discussed in later chapters.

REFERENCES

AASHTO T260-84 (1984) *Standard Method of Sampling and Testing for Total Chloride Ion in Concrete and Concrete Raw Materials*, American Association of State Highway Transportation Officers, Washington, DC.

Alldred, J.C. (1993) 'Quantifying the losses in cover-meter accuracy due to congestion of reinforcement', *Proceedings of the Fifth International Conference on Structural Faults and Repair*, Vol. 2, Engineering Technics Press, Edinburgh, pp. 125–30.

Alongi, A.A., Clemena, G.G. and Cady, P. (1993) *Condition Evaluation of Concrete Bridges Relative to Reinforcement Corrosion*, Vol. 3, *Method for Evaluating the Condition of Asphalt Covered Decks*, SHRP-S-325, Strategic Highway Research Program, National Research Council, Washington, DC.

American Concrete Institute (1990) *Corrosion of Metals in Concrete*, Report by ACI Committee 222. ACI 222R-89, American Concrete Institute, Detroit, MI.

Andrade, C., Alonso, M.C. and Gonzalez, J.A. (1990) 'An initial effort to use corrosion rate measurements for estimating rebar durability', in Berke, N.S., Chaker, V. and Whiting, D. (eds) *Corrosion Rates of Steel in Concrete*, American Society for Testing and Materials, STP 1065, Philadelphia, PA, pp. 29–37.

Andrade, C., Alonso C., Feliú, S. and González, J.A. (1995) 'Progress on design and residual life calculation with regard to rebar corrosion on reinforced concrete', in Berke, N.S., Escalante, E., Nmai, C. and Whiting, D. (eds) *Techniques to Assess the Corrosion Activity of Steel Reinforced Concrete Structures*, American Society for Testing and Materials, STP 1276, Philadelphia, PA.

ASTM C876 (1991) *Standard Test Method for Half-Cell Potentials of Uncoated Reinforcing Steel in Concrete*, American Society for Testing and Materials, Philadelphia, PA.

ASTM D1411-82 (1982) *Standard Test Methods for Water Soluble Chlorides Present as Admixes in Graded Aggregate Road Mixes*, American Society for Testing and Materials, Philadelphia, PA.

ASTM G109 (1993) *Test Method for Determining the Effects of Chemical Admixtures on the Corrosion of Embedded Steel Reinforcement in Concrete exposed to Chloride Environments*, American Society for Testing and Materials, Philadelphia, PA.

Baker, A.F. (1986) 'Potential mapping techniques', Paper 3, *Seminar on Corrosion in Concrete: Monitoring, Surveying, and Control by Cathodic Protection*, Global Corrosion Consultants, London Press Centre.

Bennett, J.E. and Mitchell, T.A. (1992) 'Reference electrodes for use with reinforced concrete structures', in *Corrosion 92 NACE*, Paper 191.

Beeby, A.W. (1985) 'Development of a corrosion cell for the study of the influences of environment and concrete properties on corrosion', *Concrete 85, Conference, Brisbane*, pp. 118–23.

Berke, N.S., Shen, D.F. and Sundberg, K.M. (1990) 'Comparison of the linear polarization resistance technique to the macrocell corrosion technique', in Berke, N.S., Chaker, V. and Whiting, D. (eds) *Corrosion Rates of Steel in Concrete*, American Society of Testing and Materials, STP 1065, Philadelphia, PA, pp. 38–51.

Broomfield, J.P., Langford, P.E. and Ewins, A.J. (1990) 'The use of a potential wheel to survey reinforced concrete structures', in Berke, N.S., Chaker, V. and Whiting, D. (eds) *Corrosion Rates of Steel in Concrete*, American Society for Testing and Materials, STP 1065, ASTM, Philadelphia, PA, pp. 157–73.

Broomfield, J.P., Langford, P.E. and McAnoy, R. (1987) 'Cathodic protection for reinforced concrete: its application to buildings and marine structures', in *Corrosion of Metals in Concrete, Proceedings of Corrosion/87 Symposium*, Paper 142, NACE, Houston, TX, pp. 222–325.

Broomfield, J.P., Rodriguez, J., Ortega, L.M. and Garcia, A.M. (1993) 'Corrosion rate measurement and life prediction for reinforced concrete structures', *Proceedings of Structural Faults and Repair – 93*, Vol. 2, Engineering Technical Press, University of Edinburgh, pp. 155–64.

Broomfield, J.P., Rodríguez, J., Ortega, L.M. and García, A.M. (1994) 'Corrosion rate measurements in reinforced concrete structures by a linear polarization device', in Weyers, R.E. (ed.) *Philip D. Cady Symposium on Corrosion of Steel in Concrete*, American Concrete Institute, Special Publication 151.

Building Research Establishment (1981) Information Paper IP6/81 *Carbonation of Concrete made with Dense Natural Aggregates*, Building Research Establishment, Garston, Watford.

Bungey, J.H. (ed.) (1993) *Non-Destructive Testing in Civil Engineering*, International Conference by The British Institute of Non-Destructive Testing, Liverpool University.

Cady, P.D. and Gannon, E.J. (1992) *Condition Evaluation of Concrete Bridges Relative to Reinforcement Corrosion*, Vol. 1, *State of the Art of Existing Methods*, National Research Council, Washington, DC, SHRP- S-330.

Clear, K.C. (1989) 'Measuring the rate of corrosion of steel in field concrete structures', *Transportation Research Record* 1211, Transportation Research Board, National Research Council, Washington, DC.

Concrete Society (1984) *Repair of Concrete Damaged by Reinforcement Corrosion*, Technical Report No. 26.

Dawson, J.L. (1983) 'Corrosion monitoring of steel in concrete' in Crane, A.P. (ed.) *Corrosion of Reinforcement in Concrete Construction*, Ellis Horwood for Society of Chemical Industry, pp. 175–92.

Feliú, S., González, J.A., Andrade, C. and Feliú, V. (1988) 'On-site determination of the polarization resistance in a reinforced concrete beam', *Corrosion*, **44**, 761–5.

Fliz, J., Sehgal, D.L., Kho, Y.-T., Sabotl, S., Pickering, H., Osseo-Assare, K. and Cady, P.D. (1992) *Condition Evaluation of Concrete Bridges Relative to Reinforcement Corrosion*, Vol. 2, *Method for Measuring the Corrosion Rate of Reinforcing Steel*, National Research Council, Washington, DC, SHRP-S-324.

Ghorbanpoor, A. and Shi, S. (1995) 'Assessment of corrosion of steel in concrete structures by magnetic based NDE techniques', in Berke, N.S., Escalante, E., Nmai, C. and Whiting, D. (eds) *Techniques to Assess the Corrosion Activity of Steel Reinforced Concrete Structures*, American Society for Testing and Materials, STP 1276, Philadelphia, PA.

Gobbett, P. (1993) 'Under constant surveillance', *Construction Maintenance and Repair*, 7 (2), 12–4. March/April 1993.

Gowers, K.R., Millard, S.G. and Gill, J.S. (1992) *Techniques for Increasing the Accuracy of Linear Polarisation Measurement in Concrete Structures*, Paper 205, NACE, Houston, TX.

Grantham, M.G. (1993) 'An automated method for the determination of chloride in hardened concrete', *Proceedings of the Fifth International Conference on Structural Faults and Repair*, Engineering Technics Press, Vol. 2, pp. 131–6.

Hausmann D.A. (1967) 'Steel corrosion in concrete: how does it occur?', *Materials Protection*, 6, 19–23.

Herald, S.E., Henry, M., Al-Qadi, I., Weyers, R.E., Feeney, M.A., Howlum, S.F. and Cady, P.D. (1992) *Condition Evaluation of Concrete Bridges Relative to Reinforcement Corrosion, Vol. 6, Method of Field Determination of Total Chloride Content*, National Research Council, Washington, DC, SHRP-S-328.

Kaetzel, L., Clifton, J., Snyder, K. and Kleiger, P. (1994) *Users Guide to the Highway Concrete (HWYCON) – Expert System*, SHRP Report and Computer Program SHRP-C-406, National Research Council, Washington, DC.

Langford, P. and Broomfield, J. (1987) 'Monitoring the corrosion of reinforcing steel', *Construction Repair*, 1 (2), 32–6.

Millard, S.G., Ghassemi, M.G. and Bungey, J.H. (1989) 'Assessing the electrical resistivity of concrete structures for corrosion durability studies', *Proceedings of UK Corrosion*, Institute of Corrosion, Blackpool, UK.

Millard, S.G. (1991) 'Reinforced concrete resistivity measurement techniques', *Proceedings of the Institution of Civil Engineers*, Part 2, **91**, 71–88.

Millard, S.G., Harrison, J.A. and Gowers, K.R. (1991) 'Practical measurement of concrete resistivity', *British Journal of NDT*, **33** (2), 59–63.

NACE (1990) Standard Recommended Practice *Cathodic Protection of Reinforcing Steel in Atmospherically Exposed Concrete Structures*, RP0290-90, National Association of Corrosion Engineers, Houston, TX.

Newman, J. (1966) *Journal of the Electrochemical Society*, **113**, 501.

Parrott L.J. (1987) *A Review of Carbonation in Reinforced Concrete*, a review carried out by C&CA under a BRE contract, British Cement Association, Slough, UK.

Rodríguez, J., Ortega, L.M. and García, A.M. (1994) 'Assessment of structural elements with corroded reinforcement', in Swamy, R.N. (ed.) *Corrosion and Corrosion Protection of Steel in Concrete*, Sheffield Academic Press, UK, pp. 171–85.

Shiessl, P. and Ruapach, M. (1993) 'Non-destructive permanent monitoring of the corrosion risk of steel in concrete', in Bungey, J.H. (ed.) *Non-Destructive Testing in Civil Engineering*, British Institute of Non-Destructive Testing, International Conference, University of Liverpool, Vol. 2, pp. 661–74.

Stark, D. (1991) *Handbook for the identification of Alkali Silica Reactivity in Highway Structures*, SHRP-C315 Strategic Highway Research Program, National Research Council, Washington, DC.

Tuutti, K. (1982) *Corrosion of Steel in Concrete*, Swedish Cement and Concrete Research Institute, Stockholm.

Titman, D.J. (1993) 'Fault detection in civil engineering structures using infrared thermography', *Proceedings of the Fifth International Conference on Structural Faults and Repair*, Engineering Technics Press, Vol. 2, pp. 137–40.

Vassie, P.R. (1991) *The Half-Cell Potential Method of Locating Corroding Reinforcement in Concrete Structures*, Transport Research Laboratory Application Guide AG9, Crowthorne, Berkshire, UK.

Whiting, D., Ost, B., Nagi M. and Cady, P. (1992) *Condition Evaluation of Concrete Bridges Relative to Reinforcement Corrosion, Vol. 5, Methods for Evaluating the Effectiveness of Penetrating Sealers*, SHRP-S-327, Strategic Highway Research Program, National Research Council, Washington, DC.

Whiting, D. and Cady, P. (1993) *Condition Evaluation of Concrete Bridges Relative to Reinforcement Corrosion*, Vol. 7, *Method for Field Measurement of Concrete Permeability*, SHRP-S-329, Strategic Highway Research Program, National Research Council, Washington, DC.

Woodward, R.J. and Williams, F.W. (1988) 'Collapse of Ynys-y-Gwas Bridge, West Glamorgan', *Proceedings of the Institution of Civil Engineers*, Part I, **84**, pp. 635–69.

5
Physical and chemical repair and rehabilitation techniques

This chapter discusses the physical and chemical options for rehabilitating corrosion damaged reinforced concrete structures. We will start by examining the conventional repairs by concrete removal and replacement. Removal techniques such as pneumatic hammers, hydrojetting and milling and their limitations are discussed. The problems of conventional repair such as continued corrosion and structural considerations are covered, along with coatings, sealers, membranes and barriers. Encasement, overlays and corrosion inhibitors are also covered in terms of corrosion performance. Mix designs and concrete chemistry will not be discussed in detail as that information is better covered elsewhere (see for instance SHRP-C-345 *Synthesis of Current and Projected Concrete Highway Technology* (Whiting et al., 1993)).

Electrochemical techniques such as cathodic protection, chloride removal and realkalization are discussed separately in Chapter 6. Rehabilitation methodology is discussed in Chapter 7 after discussion of the major rehabilitation techniques.

The appropriate repair and rehabilitation systems must be chosen for each structure according to its type, condition and future use. There is no point in spending large amounts of money on a structure due for demolition or other major works in a few years. Equally it is pointless carrying out cosmetic patching to spalled concrete on a structure expected to last another 20–50 years where future access is difficult to arrange. What looks acceptable on a bridge is not necessarily acceptable on a building, where aesthetics have an important role to play, requiring good finishes to repairs and cosmetic coatings.

One definition of the word rehabilitate is 'to restore to proper condition'. To repair is defined as 'to replace or refix parts, compensating for loss or exhaustion'. These definitions are worth bearing in mind. If we want to rehabilitate a structure we want to restore it, but not necessarily to its original condition, because if we do, it may fail again because of intrinsic flaws. We want to establish its 'proper' condition, i.e. make it resistant to corrosion. In other words, to rehabilitate the structure we may need to improve it over its original condition. To repair is merely fixing the damage. This implies that deterioration may continue. Patch repairs are just what they say. They repair the damaged concrete. They will not stop future deterioration and may, as we will see later, accelerate it. Cathodic protection and the other electrochemical techniques can rehabilitate the structure. They mitigate the corrosion process across the whole treated areas. They must be linked with compatible patch repairs. Coatings and barriers can also rehabilitate if applied well at the correct time.

It is important to repeat that we are only talking about corrosion repairs. Other problems, structural defects, alkali–silica reactivity, etc. may also need to be addressed at the same time and any rehabilitation system must be an integrated package of compatible, complementary systems. Some of the issues of compatibility will be discussed below. Specifically the incompatibility of some systems and some problems, such as ASR with desalination, will be addressed.

Whatever repair or rehabilitation option we choose, most investigations start after concrete has already been damaged by corrosion. Some concrete removal and repair is therefore required on most jobs regardless of the rehabilitation technique selected.

5.1 CONCRETE REMOVAL AND SURFACE PREPARATION

There are a number of methods for removing concrete and the choice depends on the specification, budget and contractor's preferences. If concrete is just starting to spall due to carbonation or if an electrochemical treatment like cathodic protection is planned then a simple repair of removing unsound concrete, cleaning the rebar surface, squaring the edges and putting in a sound, cementitious, non-shrinking repair material may suffice (see Figure 5.1).

Before patching carbonated concrete, the cracked and spalled concrete must be removed from around the rebar or as far as the carbonation front goes whichever is the lesser. The cementitious patch material is chosen to ensure that the steel is back in a high pH, alkaline environment. This will encourage the reformation of the passive layer to stop further corrosion.

If patching is required due to chloride corrosion then the usual specification is to remove concrete to about 25 mm behind the rebar ensuring that all corroded steel is exposed around delaminated areas and the rebar is cleaned to a near white finish to remove all rust, pits and chlorides. Figure 5.1 shows three types of repair: a bad repair, a

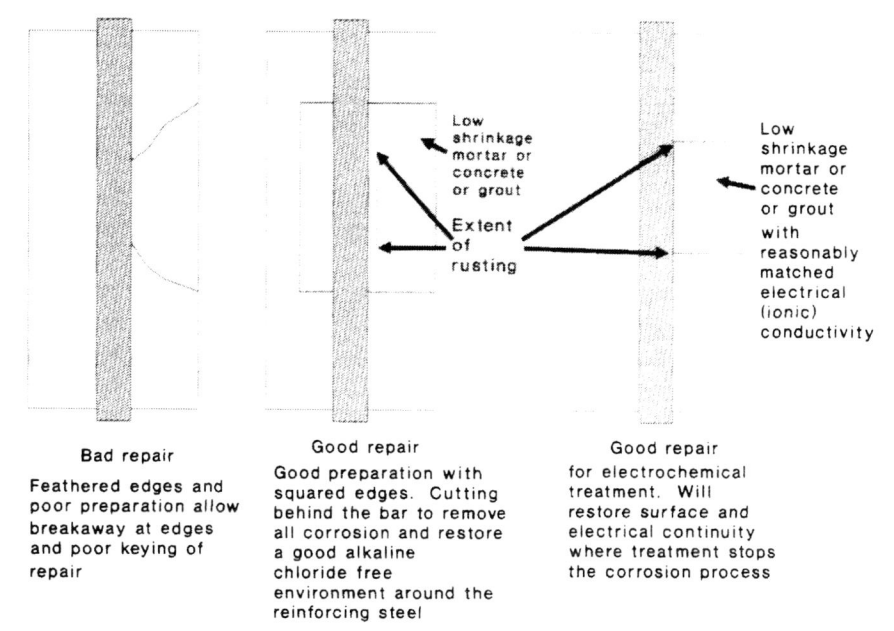

Figure 5.1 Patch repairs, bad, good and prior to electrochemical treatment.

Labels in figure:

Bad repair
Feathered edges and poor preparation allow breakaway at edges and poor keying of repair

Good repair
Good preparation with squared edges. Cutting behind the bar to remove all corrosion and restore a good alkaline chloride free environment around the reinforcing steel

Good repair for electrochemical treatment. Will restore surface and electrical continuity where treatment stops the corrosion process

Arrows indicate: Extent of rusting; Low shrinkage mortar or concrete or grout; Low shrinkage mortar or concrete or grout with reasonably matched electrical (ionic) conductivity.

good chloride repair and a patch for cathodic protection (or other electrochemical treatment). The cut edges and faces of the concrete must be square and clean of all dust and debris. The surfaces may be damped or bonding agents used to promote good adhesion between the patch repair material and the parent concrete. The choice of patch repair material is discussed below.

If an electrochemical technique is being used then repairs only need to reinstate damaged concrete. A clean rebar surface is needed and a simple patch to rebar depth in cracked, spalled and delaminated areas will prepare the structure for the application of the anode. In some cases the anode and overlay will fill the excavated areas as a single operation. Only a minimal repair is required as shown in Figure 5.1 because the treatment itself deals with the corrosion. The repair materials must be compatible with the electrochemical treatment. These issues are fully discussed in Chapter 6.

In North America some highway agencies, when repairing bridge decks without waterproof membranes, remove cover concrete in all areas where the half cell potentials exceed a threshold such as −170 mV vs. saturated calomel (−250 mV against copper/copper sulphate, CSE). It has been shown (Weyers *et al.*, 1993) that the lower the threshold the longer lasting the repair, but this must be balanced by the cost and structural implications of extensive concrete removal.

5.1.1 Pneumatic hammers

A summary of the properties and categories of hand held pneumatic breakers is given by Vorster et al. (1992). For concrete repair work, breakers usually range from about 10 to 45 kg, or a maximum of about 20 kg for vertical or overhead work. Small self-contained electric units can be used for small areas in low strength concrete but pneumatic units with a separate compressor are required for most work.

Pneumatic hammers are labour intensive but have low capital cost and are very versatile. They will cut behind the bars and get between them. Research by the Strategic Highway Research Program (SHRP) has shown that the contract terms for a repair job are a major influence upon the contractor's decision as to how concrete removal is done (Vorster et al., 1992). If concrete removal is in small areas, pneumatic hammers are more practical than the larger plant such as hydrojetting or milling equipment discussed below. If the area of concrete to be broken out is undefined at the tender stage and the contractor is paid by the square metre (or cubic metre) there is less risk with a low capital cost approach like pneumatic hammers.

Pneumatic breakers will not clean the rebars and are operator sensitive in the finish achieved. Work at the UK Transport Research Laboratory (Vassie, 1987) has shown that inadequate rebar cleaning will allow corrosion to proceed. Hydrojetting or grit blasting the rebars may be needed to remove all chloride contamination.

Typical production rates for pneumatic breakers range from about 0.025 to 0.25 m^3 per hour for breakers ranging from 10 to 45 kg. Breakers are well suited for jobs with small, discontinuous areas of concrete removal, but not for large-scale removal.

5.1.2 Hydrojetting

This is an increasingly popular method of removing concrete from decks and substructures. In theory, on a deck, a vehicle mounted system can be set up to remove concrete to a uniform depth across large areas. It will run across the deck removing unsound concrete, cleaning the rebars and leaving a surface ready for patching and overlaying. In practice concrete quality often varies and therefore the amount and depth of concrete removed will also vary. Follow up with pneumatic breakers can be required where insufficient concrete is removed.

A small 'peak' or 'shadow' often remains behind the rebar after hydrojetting (Figure 5.2). If the specification or the engineer on site requires its removal then the economic and efficiency gains of using hydrojetting can be lost. Either the client or the contractor will end up

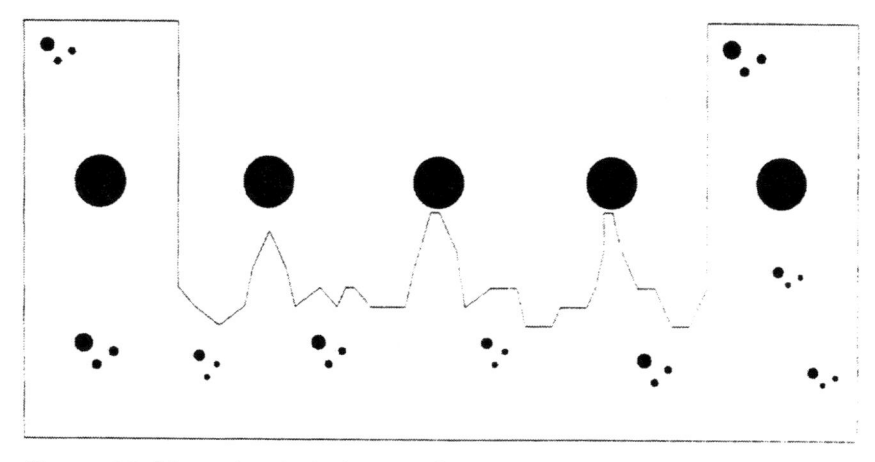

Figure 5.2 Schematic of shadowing that occurs behind rebars with hydrodemolition.

with doubled costs for concrete removal. Therefore if hydrojetting is required or preferred in a contract the contract specifications must reflect the performance of the technique. The required finish and the amount of concrete removed must reflect the performance of the hydro-

Figure 5.3 Hydrojetting a wall with a robot arm controlled system. Courtesy Professor Hans Ingvarsson, Swedish National Road Administration.

jetting technique, not a theoretical standard for preparation prior to repair.

Hydrojetting of substructures can be done manually by an operator holding a high pressure hose, or by a unit mounted on a robot arm as shown in Figure 5.3. Manual jetting requires a very high level of protection for the operator and for bystanders. Safety considerations make this approach rare in the USA.

Attention must be paid to the water run off. There is a risk of contaminating groundwater and streams with alkali, fine debris and the contaminants that have built up on the concrete surface. It may be necessary to put bunds around the repair area and collect the contaminated water or to filter the run off, collect the solid particles and prevent them from being washed away.

Hydrojetting units typically consist of an engine driving a high pressure pump connected via a high pressure hose to the nozzle which is either held manually or moved under microprocessor control within a protective guard system on a vehicle mounted system. Water is delivered at pressures ranging from about 80 to 140 MPa at flow rates ranging from 75 to 270 $l\ min^{-1}$. Hydrodemolition contractors protect information about production rates of their units. Typical rates will range from 0.25 $m^3\ hr^{-1}$ for small single pumped systems to 1 $m^3\ hr^{-1}$ for top of the range dual pumped systems. This assumes a 28 MPa concrete and removal down to 75 mm. American hydrojetting practices are discussed in Vorster *et al.* (1992). European practices are reviewed in Ingvarsson (1988) and Anon. (1993).

5.1.3 Milling machines

Milling machines can be used to remove concrete cover on decks. They must not be used right down to rebar level or they will damage the bars and the milling head. They are of limited interest in the UK and Europe for corrosion repairs on bridge decks because of the extensive use of waterproofing membranes, but are widely used in North America where heavily contaminated concrete bridge decks are frequently milled with local removal to expose corroding rebars that have delaminated the concrete.

Milling machines are far more precise than hydrojetting in removing a defined amount of cover. However, they will cause considerable damage to themselves and to the deck if they catch tie wires or stirrups, so great care must be taken to ensure that the cover is uniform and a margin is allowed to prevent exposure of the steel work. A good cover survey will ensure that as much concrete as possible is milled without risk of damage.

Removal rates can be very rapid, usually greater than 1 $m^3\ min^{-1}$

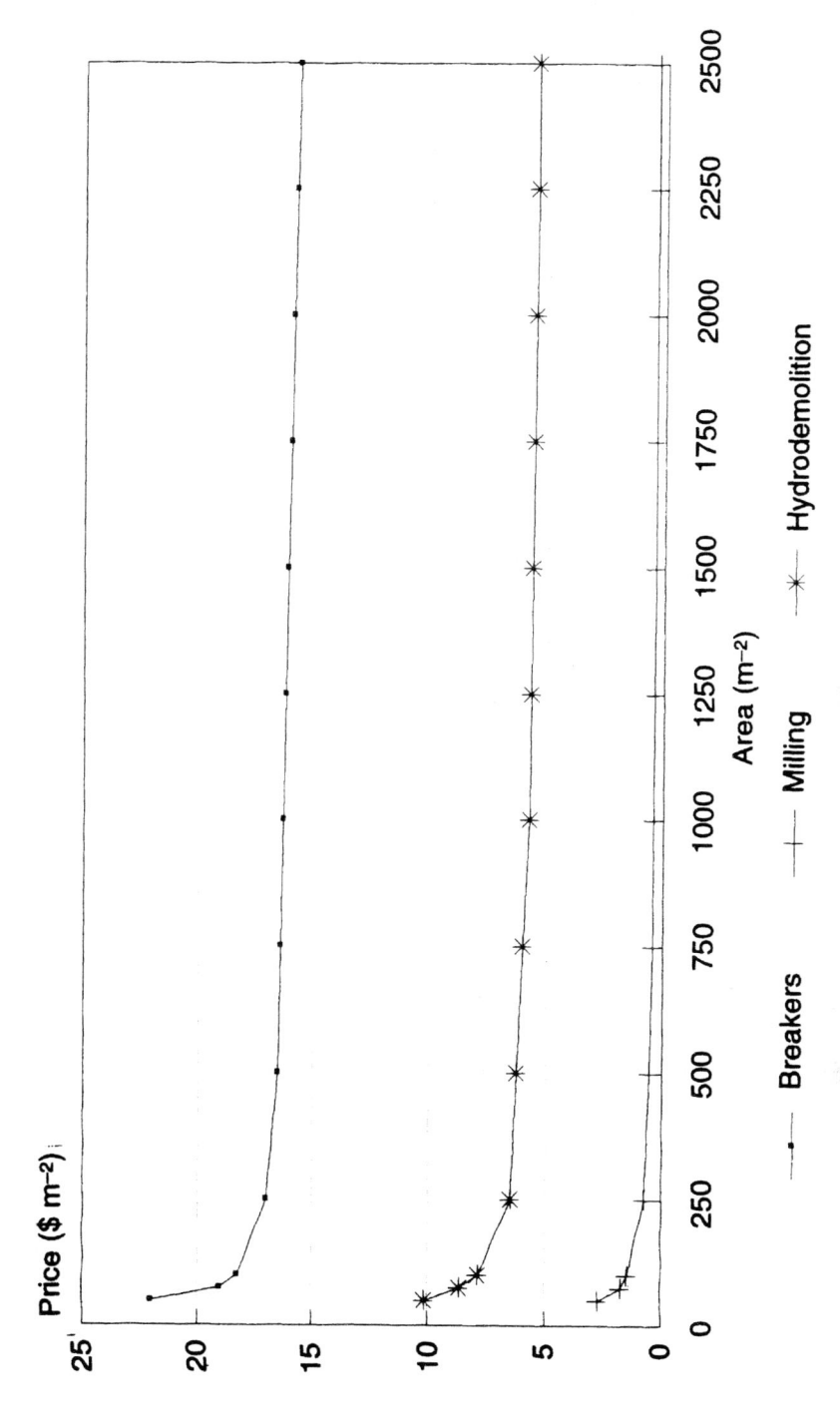

Figure 5.4 Comparisons of costs for concrete removal based on Vorster et al., (1992).

using machines with a 2 m wide cutting mandrel travelling at 1.5 m min^{-1}.

After milling, local, deeper removal with pneumatic hammers or hydrojetting is needed where cracking and spalling have occurred. Removal of concrete in areas where the steel has dropped below the corrosion threshold of −170 mV Ag/AgCl, −350 mV CSE. The deck is then patched in the delaminated areas and a dense cementitious overlay is put back on. The use of concrete overlays is discussed in Section 5.2.

5.1.4 Comparative costing

Little work has been carried out on comparative costing. This is because of the difficulty of getting comparative values, the commercial sensitivity of the rates used and the difficulty of making comparisons that are valid for more than one job.

Figure 5.4 shows a graph comparing milling, hydrojetting and jack hammering based on 1860 m^2 of concrete removal to a depth of 64 mm on a bridge deck in the USA (Vorster et al., 1992). Using this analysis breakers are never economic even for an area of about 50 m^2. This suggests that capital costs of equipment have not been compared adequately for small areas. However, it does serve to illustrate the commercial benefits of using the high tech approach.

5.1.5 Concrete damage and surface preparation

There has been some discussion in the technical literature (Vorster et al., 1992) about damage done to the parent concrete during concrete removal, especially by milling and breakers. There is no clear indication of the effect of such damage on repairs.

For the best bonding between old and new concrete a rough, clean and crack-free surface is required. The surface should be damp enough not to suck moisture out of the new concrete which is needed for hydration. However, the surface must not be soaked or ponded with water as this will increase the water:cement ratio of the new concrete, weakening it and the bond between the parent and the repair material.

The exception to this is when bonding agents are used, particularly epoxies. These must be applied according to the manufacturer's instructions, particularly where there is a requirement to place the new concrete while the bonding agent is still 'tacky'. This is often very difficult to ensure under site conditions and requires good supervision and quality assurance on site.

Hydrojetting is probably the concrete removal technique of choice for minimal damage, cleaning of rebars and removal to required depth. In all cases the edges of removed areas must be square to the surface with no 'feathering' that would be difficult to fill with reasonably sized aggregates (Figure 5.1).

5.2 PATCHES

Having removed the damaged and contaminated concrete, we must patch it. Many proprietary patch materials are on the market. The prebagged materials are most likely to be applied properly, especially to small repair areas, but they are more expensive than conventional cement/aggregate/water mixes. If repair contractors must measure quantities and mix on site, it will save money at the risk of less consistency and higher risks of shrinkage, poor bonding, etc. Specialized mix design can be carried out by concrete experts to provide pumpable, pourable and trowelable mixes. In the USA and Canada, many highway agencies have developed their own mix designs for concrete repair work for bridge decks and highway pavements based on locally available materials and the prices of additives and cement replacement materials such as micro silica, polymers or water reducing agents.

Most proprietary, prebagged mixes carry guarantees of the materials, particularly when applied as carbonation repairs. Manufacturers and applicators will be more cautious with chloride repairs as they cannot guarantee that all chloride is removed in areas adjacent to the patch. The incipient anode problem (Section 5.2.1) is far more prevalent in chloride contaminated structures as the concrete generally has lower resistivity as the chlorides are carried in by moisture. The low resistivity allows the anodes and cathodes to separate.

Fewer and fewer materials manufacturers have their own company applicators, although most have approved or recommended applicators. It is now possible to get back to back guarantees for concrete repairs to last five to fifteen years, although there is a price to pay for the premium on such insurance backed guarantees.

Some proprietary patch repair materials include bonding agents. These should be avoided in electrochemical rehabilitation work as they usually have a high electrical resistance. They may help where a high standard of workmanship is difficult to achieve, but they often require that the patch is applied at the right time when the bonding agent is ready. This is often difficult to do on site and adequate supervision is necessary to ensure that correct application is achieved.

5.2.1 Incipient anodes

We have already seen that patch repairing is not usually adequate to stop further deterioration in the presence of chloride attack. If a

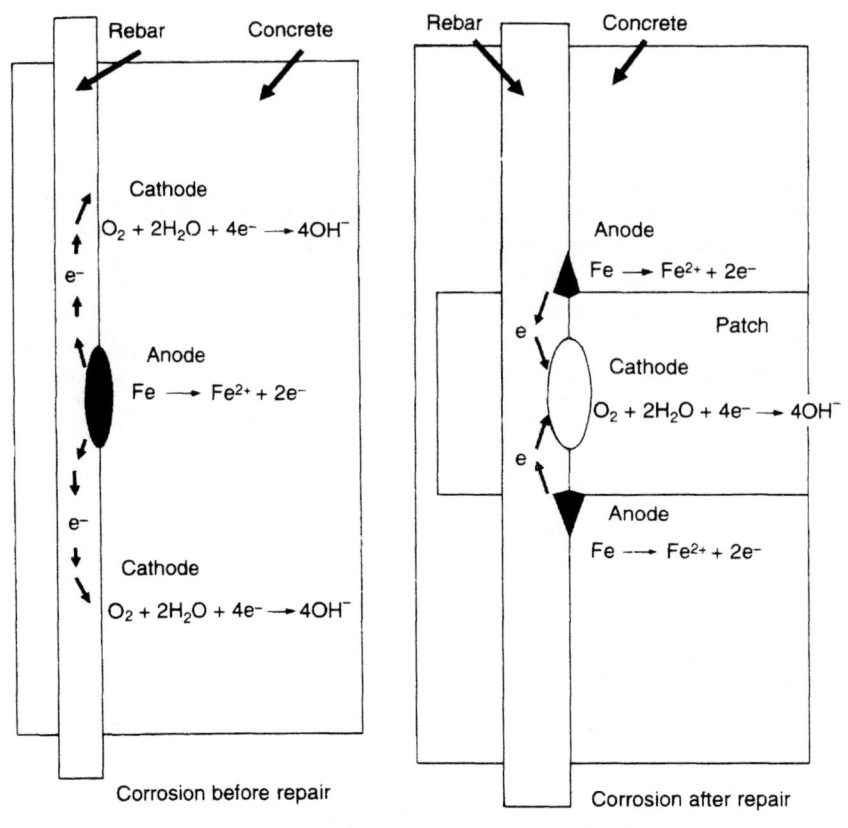

Figure 5.5 The formation of incipient anodes after patching.

structure with extensive chloride attack is to be patch repaired then it must be recognized that patching the corroding areas can accelerate corrosion elsewhere. When we stop the anodic reaction (2.1) we stop the generation of hydroxyl ions at the cathode (equation 2.2). Therefore, areas protected from corrosion because they were made cathodic by the now repaired anode will rise above the critical chloride/hydroxyl ratio and corrosion will be initiated, often around the new patch, as shown in Figure 5.5. This 'incipient anode' problem is avoided by applying an electrochemical rehabilitation technique.

One of the first cathodic protection installations in the UK was to a police station in the North of England. The author visited the station in 1987 and was shown the state of corrosion and the specification of a high quality, high cost patch repair and epoxy coating applied across the portal arches. A year later the author was invited back to examine the corrosion around the patches and it was agreed that the patches

Figure 5.6 Incipient anodes formed around the edge of a poorly applied patch on a building exposed to salt.

and the coating would be removed and a cathodic protection system would be installed. All the previous work had to be undone as the repairs were incompatible with cathodic protection. The problem was incipient anode formation around the patches. This story of incipient anodes forming around high quality patch repairs has been repeated on many structures particularly in marine and deicing salt environments. Figure 5.6 shows an example of corrosion breaking out at the edges of a patch repair which was not done with proper squared edges. This is probably partially due to incipient anode effects.

5.2.2 Load transfer and structural issues

Removing the concrete cover, either by spalling or concrete removal, redistributes the load within the structure. The exposed steel may bend once the bond between steel and concrete is lost. This can happen as a result of corrosion as well as during repair. Obviously, bending of the steel will give severe structural problems. Any significant concrete removal or corrosion damage must be assessed by a structural engineer. Propping may be necessary during repairs, particularly on substructures.

It cannot be assumed that a patch repair will take the load through it in the same way as the original concrete. This can be particularly important for slender, lightly reinforced elements on buildings. Cutting out concrete behind the rebar can be a risky operation as it leads to structural weakening and a patched structure, although it may be cosmetically attractive if a coating is applied, considerable load bearing capacity may have been lost.

The author has been involved with repairs to several structures where there was a risk of bars bending when the concrete was removed for patching due to cast in chloride induced corrosion. This led to either an inferior patch repair (because chloride laden concrete was not removed from behind the rebar) or a decision to cathodically protect the structure to avoid the need to carry out such strenuous repairs.

In some cases the need for structural support for live and dead loads and the effort to transfer loads into the repair (for instance as done as an exercise on the Midland Links elevated motorway sections in the UK) have made such repairs completely uneconomic. This is one reason for choosing an electrochemical rehabilitation technique such as cathodic protection, as only 'cosmetic' repairs are required and the minimum of concrete is removed.

5.3 COATINGS, SEALERS, MEMBRANES AND BARRIERS

One of the attractions of construction in concrete is that coatings are not usually required so that maintenance costs are lower than, for instance, when using structural steel. Many of those associated with concrete structures are reluctant to apply coatings. However, coatings can be beneficial in excluding undesirable species such as chlorides and carbon dioxide, or cosmetically restoring the appearance after concrete repair. Even the most carefully matched repair will weather differently from the original material, eventually requiring a coating to hide it again.

A huge range of coatings and sealers can be applied to concrete. Apart from looking different from each other, they also perform different tasks. There are anti-carbonation coatings which should be applied after carbonation repairs to stop further carbon dioxide ingress. These can be of a variety of different generic types. The carbonation resistance should always be checked rather than relying upon coating formulation as testing has shown that not all acrylics (for instance) have good carbonation resistance (Robinson, 1986). Penetrating sealers, silanes, siloxanes, siloxysilanes, etc. are recommended to reduce chloride ingress. Penetrating sealers have been known to accelerate carbonation in laboratory tests so care should be taken in deciding where and when they are applied.

Once chloride induced corrosion has started it is very unlikely that a coating of any sort will stop it. There is sufficient moisture and oxygen within the concrete to generate the small amount of rust needed to crack the concrete and then the water and oxygen will have a free path to the steel surface. Once the steel is depassivated a coating will not restore its passivity.

It is also unlikely that a coating applied in the field to a real structure will seal it totally. When chlorides get into concrete they form hygroscopic salts. These will condense water from the atmosphere and from elsewhere in the concrete. It should be noted that the 'average' humidity in the UK and much of the US east coast is around 70–80%. This is close to the optimum for corrosion. Any sealing coat may actually help to retain the optimum humidity level in the concrete to enhance corrosion.

The total or near total sealing of concrete can cause problems. Concrete is porous and contains water and oxygen in its pores. The amount of water as vapour and as liquid in the pores is in equilibrium with the atmosphere. If a coating seals water in the pores then when the atmospheric humidity drops or the temperature increases very large local forces can be exerted against the barrier coating causing blistering and coating failure.

5.3.1 Carbonation repairs

If a combination of patching and sealing is required for a carbonated structure, the patch repairs must remove all carbonated concrete around the steel. In some cases this will only require exposure of the front face but in more extreme conditions it may need concrete removal behind the steel to restore the passive alkaline environment. Much has been published about anti-carbonation coatings and their effectiveness but it is not easy to find independent information about specific products (Robinson, 1986). Comparison of product information may be required. This may include the requirement for independent testing. The principle of anti-carbonation coatings is that they are porous enough to let water vapour move in and out of the concrete but the pores are too small for the large carbon dioxide molecule to pass through.

5.3.2 Coatings against chlorides, penetrating sealers

Penetrating sealers have been recommended as a way of stopping chlorides getting into concrete. The chemistry of the process is that silanes, siloxysilanes and similar chemicals will penetrate the pores of the concrete and react with the water in the pores to form a hydrophobic (water repelling) layer that stops water getting in as a liquid (that may carry salt with it), but allows water vapour in and out of the concrete so that it will 'breathe'.

The penetrating sealer is within the concrete so it is protected from physical damage and degradation by ultraviolet light, etc. The problem is that the colourless liquid must be applied so that most of it gets into the pores and there is enough water in the pores to react, but not too much which will wash the sealer back out so it will not coat the pores. The time between applications and the permissible weather conditions can turn an apparently inexpensive protection system into a very expensive one. There is still some debate over the true penetration depth of sealers to good quality concrete in the field.

Since 1986 pure (100%) isobutyltrimethoxysilane has been specified to prevent chloride ingress on exposed concrete on UK highway bridges. A penetration depth of 2–4 mm has been claimed, but this may reduce to 1 mm or less in new, well cured concrete with a low water/cement ratio. Other highway agencies have used different formulations with some siloxanes in solutions and larger molecules that are less volatile. All these chemicals form silicone resins within the pores. The larger molecules are less volatile and therefore easier to get to the concrete surface, but once there they are less mobile and penetrating.

In the USA many highway agencies have applied penetrating sealers to concrete bridge decks. No definitive research has shown whether sealers work on trafficked surfaces, or for how long they are effective.

The UK Highways Agency now specifies penetrating sealers for the concrete surfaces of bridge substructures exposed to chlorides. The German Department of Transport has done so for several years. This is an inexpensive method of slowing the ingress of chlorides before corrosion starts, subject to the application problems discussed above. There have been discussions in the trade press about what materials should be used and how they should be applied. Work is now proceeding on developing a performance specification for penetrating sealers.

In the USA the SHRP program developed two field tests for penetrating sealers (Whiting et al., 1992). These can be used *in situ* to determine the effectiveness of sealers. One is a surface conductivity test and the other is a variation on the surface absorption test (ISAT). If silane based sealers are being used then these tests are recommended. They can be used to rank the performance of different materials on a given structure. Once the structure is coated, they will measure the effectiveness of the application and check uniformity across the structure.

The alternative is to take cores and do laboratory tests of chloride permeability or examine for silane penetration. This is slow, expensive and damaging to the structure.

5.3.3 Waterproofing membranes

According to a report by the Organization for Economic Cooperation and Development (1989), most European countries put waterproofing membranes on their decks where there is a risk of chloride ingress. This is usually a sheet, a spray or 'squeegee' applied liquid system applied over the new concrete surface, sometimes with a base or primer coat and with protective layers, with a final asphalt overlay. A synthesis of practice of waterproofing membranes for bridge decks was recently published by Manning (1995). Approximately 50 systems were tested by the UK Transport Research Laboratory in a trial of membrane systems (Price, 1989). Figure 5.7 shows a schematic of the components

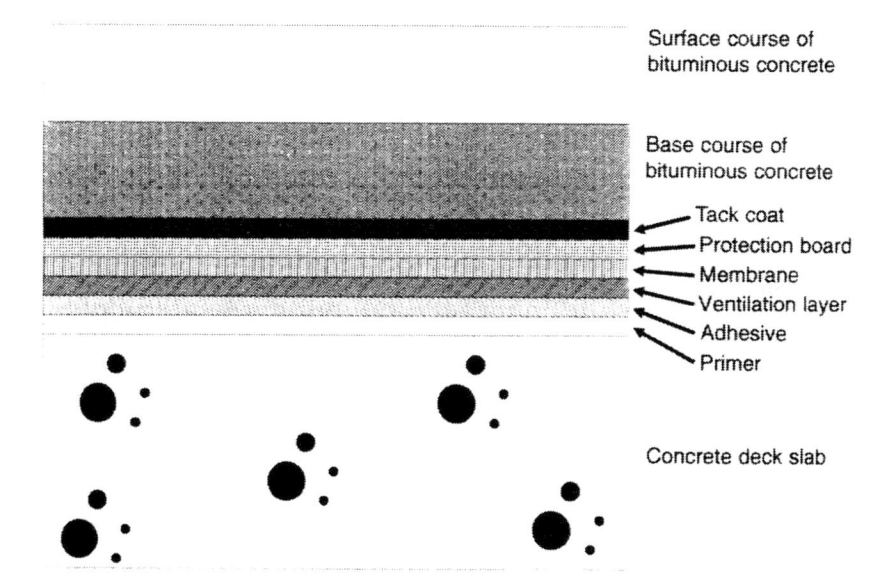

Figure 5.7 Schematic of possible components of a waterproofing system from Manning (1995).

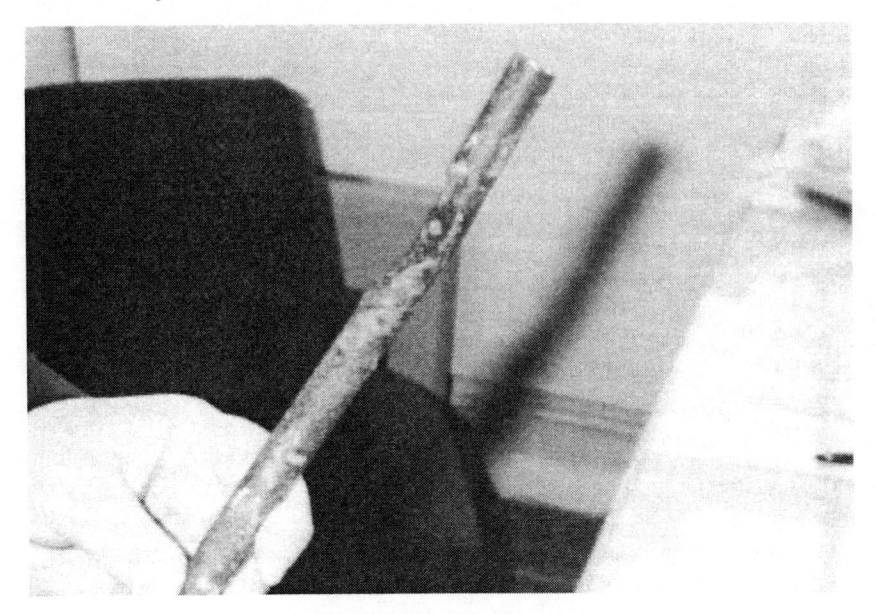

Figure 5.8 When membranes fail and water gets underneath local corrosion can be severe and lead to heavy loss of section without cracking and spalling the concrete.

of a waterproofing system. Figure 5.8 shows one of the membranes after testing.

These systems have been of variable quality in the past, although tighter specifications have been introduced by most national DOTs. Membranes have failed at joints, curbs and drains where chloride laden water could get under them. Some were damaged or destroyed by the hot application of the asphalt wearing course over them. A survey of site practice and failures in the UK is given in Price (1991).

However, the lack of waterproof membranes is why most North American cathodic protection systems are on bridge decks, where membranes were not used, while most European cathodic protection systems applied to bridges are on the support structure. Membranes and cathodic protection are not easily compatible as gases are evolved by cathodic protection systems which could be trapped by a waterproofing system. This is discussed in the next chapter.

The problem with waterproofing membranes is that they have a 10- to 15-year lifetime. This means that they must be replaced and any areas of failure repaired. There is also a problem with severe pitting and 'black rust' as discussed in Chapter 2. This is observed at the weak points listed above, particularly at joints. This can give rise to structural concerns at half joints where the steel is part of the main support

Figure 5.9 Waterproofing membranes can fail to bond, dissolve into the concrete under the application of the hot mix for the wearing course, or be punctured by the aggregates. Tighter specifications are now used in the UK to avoid systems prone to such failures. This photograph shows some of the membranes after testing at the Transport Research Laboratory.

system. An example of severe loss of section seen underneath membranes is shown in Figure 5.9.

Waterproofing membranes have been developed for car park decks that can be applied without asphalt overlays. This minimizes the load on the deck and the reduction of headroom which an asphalt overlay would create. These overlays are able to resist traffic loads from cars and small vehicles.

5.3.4 Barriers and deflection systems

These are the logical extension to the waterproofing membrane. Often one of the simplest ways of reducing the deterioration rate due to chloride attack is simple deflection of chloride laden water away from the concrete surface. This can sometimes be done with the introduction of guttering and drainage on buildings or bridge substructures subject to salt water run off. This can be a very cost-effective way of at least stopping the acceleration of decay. It can also extend the life of rehabilitation systems such as patches, cathodic protection anodes and chloride removal treatments.

A more expensive barrier approach can be seen on bridges where the reinforced concrete has been clad in a brick or masonry finish, usually for cosmetic reasons. The salt rarely penetrates to the concrete surface. However, this option is rarely available for reasons of cost.

This approach is not effective once the critical chloride level has been exceeded. It may be necessary to calculate the likely build up of chloride at rebar depth if applying barriers or coatings after chlorides have built up in the concrete cover. This is required even if critical chlorides have not reached rebar depth as they will continue to diffuse in after the chloride ingress is curtailed by coatings or barriers.

5.4 ENCASEMENT AND OVERLAYS

After milling, and deeper removal where cracking and spalling have occurred, a bridge or car park deck can then patched in the delaminated areas and a dense cementitious overlay of micro silica, polymer modified or low slump, low water/cement ratio concrete or other suitable waterproofing (chloride ingress resistant) deck concrete is put back on. This will slow the corrosion rate and the appearance of further delaminations (Figure 5.10).

These techniques are popular methods in the USA for delaying, hiding or preventing chloride induced delamination on substructures

Figure 5.10 Auger distribution of a low slump dense (Iowa) concrete overlay. Courtesy Mr Bernard Brown, Iowa DoT.

and bare concrete bridge decks. Florida DOT has found that encasing bridge columns in concrete is not effective in stopping corrosion on the columns of the bridges around the Florida coast. However, several DOTs in northern states believe that they are having some success. Their exposure to chlorides, moisture and high temperature (hence high corrosion rates) is far less severe than in Florida.

Overlays or encasement may absorb some of the chlorides, reducing the level at the concrete surface. They will certainly reduce the high chloride gradient that drives chloride further into the concrete if they are coupled with removal of some or all of the old cover concrete.

Recent research into overlays shows that many of them suffer from shrinkage cracks but still seem to be effective in stopping or slowing corrosion. Overlays last longest in the more southerly states like Virginia when compared with the more northerly ones like New York, that suffer from colder winters, heavier salt application and also higher traffic density.

Overlays may be of polymer modified concretes, low slump dense concrete (the Iowa mixes), or micro silica concretes. They vary in cost and ease of application. State highway agencies generally develop one or two mix designs that suit their purposes and make competitive bidding possible.

Encasement on substructure columns is less routine than overlay application on decks. The concrete is broken out where it is damaged and an oversized shutter is applied. Concrete is then pumped or placed into the shutter enlarging either the whole column or the damaged section. This technique is extensively applied on structures such as wharves and bridge substructures in marine environments.

5.5 SPRAYED CONCRETE

Sprayed concrete is a rapid method of applying concrete to soffits or vertical surfaces. It can be used over patches and to overlay metal mesh anodes for cathodic protection as described in Section 6.6. It is sometimes applied as a temporary 'cosmetic' repair in the northern USA and Canada when concrete has spalled or is in danger of spalling. As it does nothing to slow the corrosion rate, a sprayed concrete repair is comparable in effectiveness to patch repairing. Their effectiveness is highly dependent on the amount of concrete removed. Like patching and overlaying they may suffer from the same 'incipient anode' problem.

Where the concrete is applied as shotcrete, water and dry mix cement and aggregates are sprayed through two nozzles and mix 'in flight' and on impact. Considerable operator skill is required to apply shotcrete effectively with a minimum of delaminations. This is particularly true when cathodic protection anodes are being installed. In the case of cathodic protection, the shotcrete is required to stick to the original finished surface which may not be as rough or absorptive as the surface exposed by concrete removal from corrosion damaged areas.

Inadequate mixing or inappropriate quantities at the nozzle can lead to voids, honeycombing or lenses of unhydrated concrete. Extensive trials should be undertaken to show that good bond and mixing is achieved on site. Thorough inspection afterwards should show a high consistency of application achieved and maintained throughout the application.

Wet mix spraying of mortars is becoming more popular as proprietary mixes have been developed. The consistency of the mix is easier to control in wet mix application if segregation does not occur in the mixer before spraying. It is often possible to finish the surface with wet sprayed mix which is not possible with dry mixed shotcrete.

5.6 CORROSION INHIBITORS

The ability of nitrites to stop corrosion when added to concrete has been known for many years and a proprietary calcium nitrite additive has been used in a number of structures including bridges and car parks. Florida DOT is carrying out trials of the use of calcium nitrite in their bridge substructures to prevent marine corrosion in the bridges on the Florida coast.

Recently, the idea of applying inhibitors to stop corrosion in existing structures has received some attention from researchers and the producers of corrosion inhibitors. The principle of most inhibitors is to develop a very thin chemical layer usually one or two molecules thick, on the steel surface, that inhibits the corrosion attack. Inhibitors can prevent the cathodic reaction, the anodic reaction or both (cathodic, anodic and ambiodic inhibitors). They are consumed and will only work up to a given level of attack (i.e. chloride content). The concern with inhibitors is that they will suppress the generalized corrosion; but if the amount available is inadequate due to low dosage or consumption, there could be localized, pitting attack. This phenomenon has not been clearly demonstrated in concrete but can occur in theory and has been observed in other corrosion environments.

There are now a number of inhibitors being offered to improve the effectiveness of concrete repairs. They can be applied as coatings on the surface or on to the exposed steel at patch repairs, incorporated into the patch repairs, applied in grooves or drilled holes in the concrete cover or incorporated into concrete overlays.

Calcium nitrite has the advantage that it will not affect the concrete significantly if added to a repair mix, or applied to a surface and the mix applied over it. However, calcium nitrite is a set accelerator so care must be taken that 'flash setting' does not occur. Otherwise it is straightforward to use, other than some concern about the freeze–thaw resistance. Adequate air entrainment is advised for mixes including calcium nitrite.

Vapour phase inhibitors (VPIs) are volatile compounds that can be incorporated into a number of carriers such as waxes, gels and oils. They will diffuse through the air, or the concrete pores, to the steel surface. In principle their ability to diffuse as a vapour gives them an advantage over liquid inhibitors. However, they can also diffuse out of the concrete unless trapped in place. They may diffuse poorly through saturated concrete.

There are other materials on the market that have different effects on the steel or the concrete to enhance the alkalinity, block the chloride and reduce the corrosion rate. Some are true corrosion inhibitors, some are hybrid inhibitors, pore blockers and alkali generators.

Although some field trials have been underway for several years, it is difficult to show how effective any of these systems are in suppressing corrosion in existing structures. Diffusion of the chloride ion into concrete is slow and most of the liquid inhibitor molecules are larger than the chloride ion so we would expect them to be even slower to reach the rebar from the surface. Many evaluations of inhibitors have been simplistic, not fully representing field concretes or conditions, or are in early stages with no results properly reported. The true effect of an inhibitor can only be assessed by corrosion rate measurement before and after application and with regular monitoring of a treated area and a control area. Such evaluations are in their earliest stages.

Several methods of speeding the inhibitor to the steel surface have been proposed. Drilling or grooving concrete to apply inhibitors is expensive and damaging. Simple spraying on to a dry surface will probably be very dependent upon the concrete quality and the moisture level in the concrete.

A systematic evaluation of corrosion inhibitors is now underway at the UK Building Research Establishment; preliminary results may be available in 1997–8. SHRP undertook field trials of two inhibitor applications on bridge decks and substructures in the USA. These will be monitored from 1994 to 1999. Other trials are underway in Europe on specific materials.

The advantage of inhibitor application is that they can reduce the 'incipient anode effect' around patches if they are incorporated into the repair mix and then diffuse into the concrete. Their disadvantage is the cost of the materials, difficulty of application and unknown lifetime and effectiveness. However, with proper evaluation and monitoring, we may learn a lot more about these materials over the next few years.

REFERENCES

Anon. (1993) 'Hydro-demolition', *Construction Repair*, **7** (2), 32–6.

Concrete Society (1991) *Patch Repair of Reinforced Concrete: Subject to Reinforcement Corrosion*, Technical Report No. 38.

Ingvarsson, H. and Eriksson, B. (1988) 'Hydro demolition for bridge repairs', *Nordisk Betong*, **2–3**, 49–54.

Manning, D.G. (1995) *Waterproofing Membranes for Concrete Bridge Decks: A Synthesis of Highway Practice*, NCHRP Synthesis 220, National Cooperative Highway Research Program, Transportation Research Board, National Research Council, Washington, DC.

Organization for Economic Cooperation and Development (1989) *Durability of Concrete Road Bridges*, OECD, Paris.

Price, A.R.C. (1989) *A Field Trial of Waterproofing Systems for Concrete Bridge Decks*, TRRL Research Report 185, Transport Research Laboratory, Crowthorne, Berkshire, UK.

Price, A.R.C. (1991) *Waterproofing of Concrete Bridge Decks: Site Practice and Failures*, TRRL Research Report 317, Transport Research Laboratory, Crowthorne, Berkshire, UK.

Robinson, H.L. (1986) 'Evaluation of coatings as carbonation barriers', *Proceedings of the Second International Colloquium on Materials Science and Restoration*, Technische Akademie Essingen, Germany.

Vassie, P.R. (1987) *Durability of Concrete Repairs: The Effect of Steel Cleaning Procedures*, Research Report 109, Transport Research Laboratory, Crowthorne, Berkshire, UK.

Vorster, M., Merrigan, J.P., Lewis, R.W. and Weyers, R.E. (1992) *Techniques for Concrete Removal and Bar Cleaning on Bridge Rehabilitation Projects*, SHRP-S-336, National Research Council, Washington, DC.

Weyers, R.E., Prowell, B.D., Sprinkel, M.M. and Vorster, M. (1993) *Concrete Bridge Protection, Repair and Rehabilitation Relative to Reinforcement Corrosion: A Methods Application Manual*, Strategic Highway Research Program Report SHRP-S-327, National Research Council, Washington, DC.

Whiting, D., Ost, B. and Nagi, M. (1992) *Condition Evaluation of Concrete Bridges Relative to Reinforcement Corrosion*, Vol. 5, *Methods for Evaluating the Effectiveness of Penetrating Sealers*, Strategic Highway Research Program Report SHRP-S-327, National Research Council, Washington, DC.

Whiting, D., Todres, A., Nagi, M., Yu, T., Peshkin, D., Darter, M., Holm, J., Andersen, M. and Geiker, M. (1993) *Synthesis of Current and Projected Concrete Highway Technology*, Strategic Highway Research Program Report SHRP-C-345, National Research Council, Washington, DC.

6

Electrochemical repair techniques

6.1 BASIC PRINCIPLES OF ELECTROCHEMICAL TECHNIQUES

As we saw in reactions (2.1) and (2.2), corrosion occurs by the movement of electrical charge from an anode (a positively charged area of steel where steel is dissolving) to the cathode (a negatively charged area of steel where a charge balancing reaction occurs turning oxygen and water into hydroxyl ions). This means that the process is both electrical and chemical, i.e. electrochemical. We have also seen that, in the case of chloride attack, patch repairs are only a local solution to corrosion and repairing an anode may accelerate corrosion in adjoining areas due to the incipient anode effect (Section 5.2.1 and Figure 5.5).

One solution to this problem is to apply an electrochemical treatment which will suppress corrosion across the whole of the treated structure, element or area treated. Figure 6.1 shows the basic components of an electrochemical treatment system. They are a DC power supply and control system, and an anode (temporary or permanent), usually distributed across the surface of the concrete. Electrochemical methods work by applying an external anode and passing current from it to the steel so that all of the steel is made into a cathode. Three techniques are described here. The best known and most established technique is cathodic protection. A newer alternative for chloride contaminated structures is chloride extraction (also known as electrochemical chloride removal and as desalination). A method for treating carbonated concrete has also been developed and is gaining rapid acceptance as a rehabilitation for carbonation in buildings and other structures. This is known as realkalization. The principal differences between the processes are that cathodic protection is a permanent installation, while the other two are temporary, taking about one to eight weeks to treat the reinforced concrete. This chapter will discuss the electrochemical

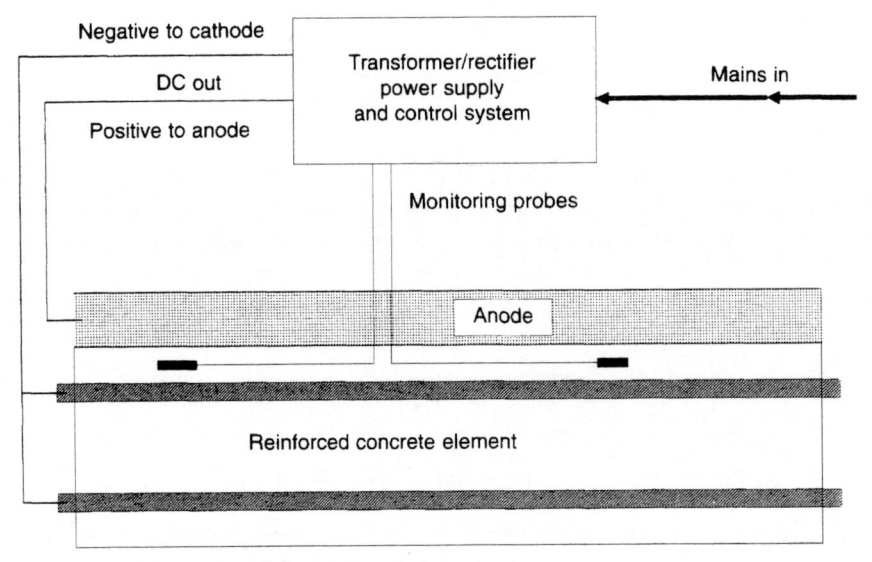

Figure 6.1 Schematic of electrochemical protection.

techniques available to engineers who have to protect corroding reinforced concrete structures.

All of these treatments are suitable for treating reinforcing steel in concrete. However, the passing of high levels of electric current can have adverse effects on concrete and on steel. They can only be applied with great care to structures containing prestressing steel (pre- or post-tensioned) and to structures suffering from alkali silica reactivity (ASR). They cannot be applied through electrically insulating layers or patches. These problems are discussed below.

6.2 CATHODIC PROTECTION

6.2.1 Theory and principles of impressed current systems

Impressed current cathodic protection works by passing a small direct current (DC) from a permanent anode fixed on to the surface or into the concrete to the reinforcement. The power supply passes sufficient current from the anode to the reinforcing steel to force the anode reaction (6.1) to stop:

$$Fe \rightarrow Fe^{2+} + 2e^- \quad (6.1)$$

and make the cathodic reaction (6.2) the only one to occur on the steel surface. The cathodic reaction will then occur across the reinforcement network:

$$H_2O + \tfrac{1}{2}O_2 + 2e^- \rightarrow 2OH^- \tag{6.2}$$

However, another reaction can occur if the potential gets too negative:

$$H_2O + e^- \rightarrow H + OH^- \tag{6.3}$$

This hydrogen evolution reaction can lead to hydrogen embrittlement of the reinforcing steel. Monatomic hydrogen can diffuse into the steel and get trapped at grain boundaries or other defects in the crystalline matrix of the steel, weakening it and causing failure under load. This problem is negligible for normal reinforcing steel but is of considerable concern for prestressed structures where the high tensile steel can be very susceptible to hydrogen embrittlement. The steel is loaded to up to 75% of its ultimate tensile strength by the prestressing process. This means that it is liable to catastrophic failure.

The problems of hydrogen embrittlement and of gas evolution are usually controlled by limiting the potential of the steel to below the hydrogen evolution potential. However, in acidic pits or crevices it may be possible for the potential to exceed the hydrogen evolution potential without being sensed by measuring electrodes. The cathodic protection of prestressed structures should only be undertaken with great care and input from experienced corrosion experts.

The generation of hydroxyl ions in equations 6.2 and 6.3 will increase the alkalinity and help to rebuild the passive layer where it has been broken down by the chloride attack. The chloride ion itself is negative and will be repelled by the negatively charged cathode (reinforcing steel). It will move towards the (new external) anode. With the carbon based anodes it may then combine to form chlorine gas at the anode:

$$2Cl^- \rightarrow Cl_{2(gas)} + 2e^- \tag{6.4}$$

The other major reaction at the anode is the formation of oxygen:

$$2OH^- \rightarrow H_2O + \tfrac{1}{2}O_2 + 2e^- \tag{6.5}$$

and

$$H_2O \rightarrow \tfrac{1}{2}O_2 + 2H^+ + 2e^- \tag{6.6}$$

Reaction 6.5 is the reverse of reaction 6.2, i.e. alkalinity is formed at the cathode (enhancing the passivity of the steel) and consumed at the anode (leading to acid etching of the concrete). These and related reactions can carbonate the area around the anode (especially where carbon based anodes are used, where the carbon also turns into carbon dioxide) and can lead to attack of the cement paste once the alkalinity is consumed.

We can therefore see that three factors must be taken into account when controlling a cathodic protection system:

1. There must be sufficient current flow to overwhelm the anodic reactions and stop, or severely reduce the corrosion rate.
2. The current must stay as low as possible to minimize the acidification around the anode (and the attack of the anode for those that are consumed by the anodic reactions).
3. The steel should not exceed the hydrogen evolution potential, especially for prestressed steel, to avoid hydrogen embrittlement.

The balancing of these requirements will be discussed below under criteria for control of impressed current cathodic protection systems.

One of the more confusing facts of cathodic protection is that when we carry out a half cell potential survey of a reinforced concrete structure, the most negative areas are those that are corroding most; the

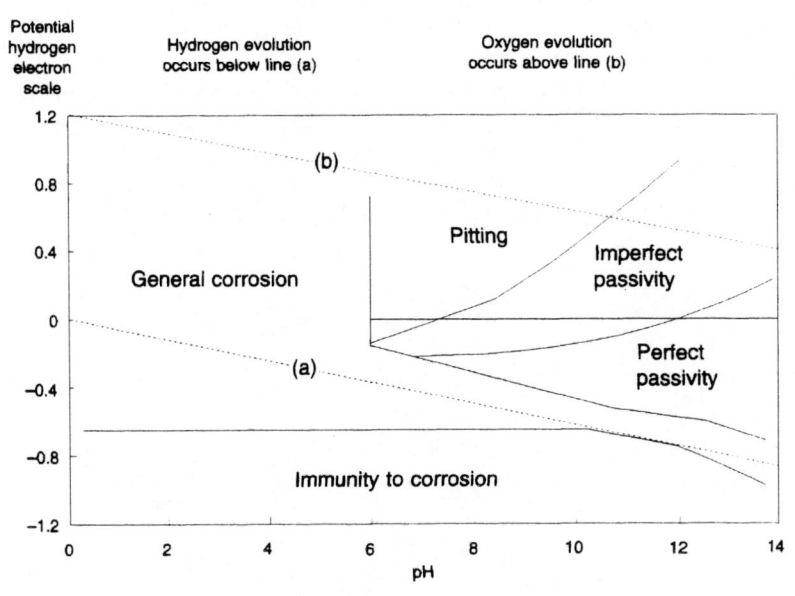

Figure 6.2 Pourbaix diagram showing experimental conditions of immunity, general corrosion, perfect and imperfect passivity for iron in solutions containing 355 ppm chloride. From Pourbaix (1973).

more positive areas are not corroding, or are cathodic. However, when we apply cathodic protection we make the reinforcing steel more negative, not less. This is because we are using a new site for the anodic reaction and are injecting electrons into the steel from outside.

One definition of effective cathodic protection is to depress the potential of the cathodes to the level of the anodes, thus stopping current from flowing between anodic and cathodic areas (Mears and Brown, 1938). This works because cathodes are more easily polarized (potential shifted) than anodes. We saw this phenomenon in Section 4.11 where the effect of an external current on the half cell potential allows us to calculate the corrosion rate.

Another way of looking at the theory of cathodic protection is to look at the Pourbaix diagram (Pourbaix, 1973; Morgan, 1990) for iron in chloride solution. This shows that there are conditions where steel corrodes, and areas where protective oxides form and an area of immunity to corrosion depending upon the pH and the potential of the steel. Ideally we would like to depress the potential sufficiently to reach the immune zone (Figure 6.2). In practice we do not do that for reasons described in the Section 6.5 on control criteria.

6.2.2 Theory and principles of sacrificial anode systems

There are two forms of cathodic protection, impressed current and sacrificial. The impressed current system has been described above and is the system conventionally used for atmospherically exposed reinforced concrete structures. An alternative method is to directly connect the steel to a sacrificial or galvanic anode such as zinc without using a power supply. This anode corrodes preferentially, liberating electrons with the same effect as the impressed current system, e.g.

$$Zn \rightarrow Zn^{2+} + 2e^-$$

The system is illustrated schematically in Figure 6.3. It should be noted that the potential of the steel is not uniform, but if the system is working properly it will all be more negative than the anode with respect to a standard half cell.

The same phenomenon is used in galvanizing where a coating of zinc is applied over steel to corrode preferentially, protecting the steel (hence the alternative name of galvanic cathodic protection). However, the main restriction on this system is that the zinc has only a small driving voltage when coupled to steel. This is about 1000 mV with respect to passive steel and 600 mV for actively corroding steel. While a galvanizing system puts the two metals in direct contact, with sacrificial anode cathodic protection (SACP) there in an electrolyte to carry the

112 *Electrochemical repair techniques*

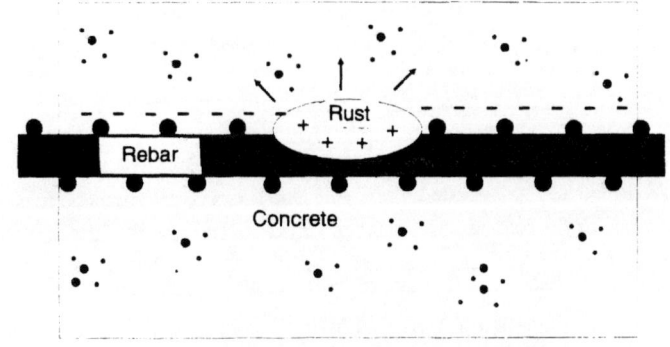

Corrosion proceeds by forming positive and negative
areas on the steel (anodes and cathodes)

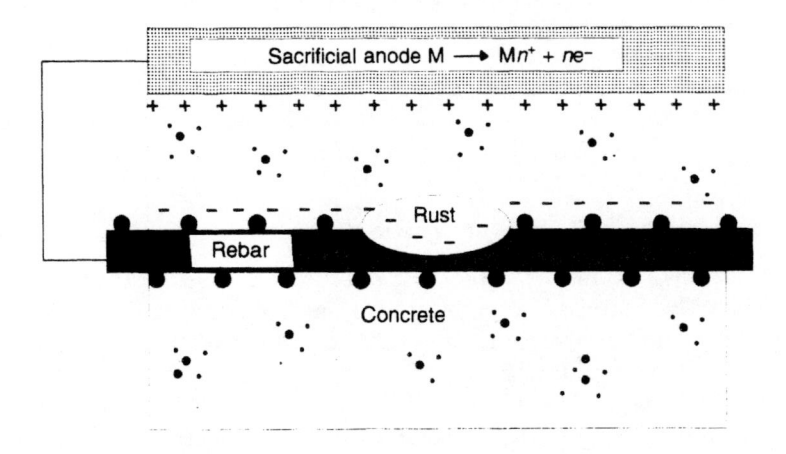

Sacrificial anode M $\longrightarrow$ Mn^+ + ne^-

Sacrificial Cathodic Protection makes all the steel
negative by dissolution of anode metal to generate electrons
Resistance must be low for enough current to pass

Figure 6.3 Schematic of sacrificial anode cathodic protection. This makes all the steel negative by dissolution of anode metal to generate electrons. Resistance must be low for enough current to pass.

Figure 6.4(a) and **(b)** Sacrificial anode cathodic protection using expanded aluminium mesh with a concrete overlay (not yet applied). June 1995. Acknowledgements to the US Federal Highways Administration.

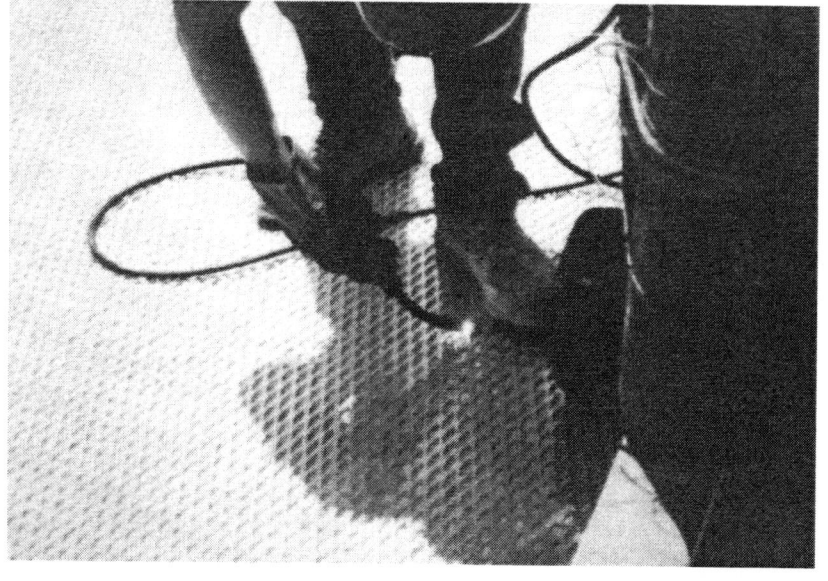

Cathodic protection

current. The resistance of the electrolyte is crucial to the performance of the system. The SACP system is still largely experimental except in Florida where it has been used extensively on prestressed concrete piles in the sea (Kessler et al., 1995). SACP is used because the risk of hydrogen embrittlement of the prestressing is negligible and concrete resistivity is low due to the marine exposure conditions (Hartt et al., 1994).

The resistivity of normal 'inland' concrete (i.e. concrete not exposed to continuous wave action and sea spray) is high compared to concrete continuously exposed to sea water and other aqueous, non-cementitious electrolytes where zinc sacrificial anodes are often used. The lower resistivity is essential for effective SACP. This resistivity can be made higher by the formation of oxides as the sacrificial anode corrodes. Unlike a free flowing electrolyte, the corrosion products are not washed away in concrete. These problems mean that sacrificial anode cathodic protection is still experimental in atmospherically exposed concrete structures, although it is being experimented with on bridge decks (Whiting et al., 1995) and on marine bridge substructures outside Florida. Applications and experiments with these systems are discussed below. SACP systems are routinely used on North Sea oil platforms on concrete and steel structures.

There are a number of elements and their alloys which are more active than steel in the electrochemical series. The main practical metals are zinc, aluminium and magnesium. All of these metals are used, often in alloyed form, as sacrificial anodes for steel structures in water. Zinc and aluminium alloys are being used in experimental SACP systems (Whiting et al., 1995). An experimental system using an expanded aluminium mesh anode is shown in Figure 6.4.

Although SACP systems may be limited in their application due to the limited voltage generated by sacrificial anodes compared with an impressed current system, they have advantages. The principal advantage is the lack of a power supply. This makes the system cheaper to build and easier to run. The fact that the current and voltage cannot be regulated means that monitoring requirements are minimal. The anode can be directly shorted to the cathode, with no external wiring.

In experimental systems installed on bridges in the USA, sections of rebar have been isolated from the network and connected via an ammeter to the rest of the steel. This allows current flows into a particular area to be monitored. Similarly, an area of anode has been isolated and current flow measured to determine how much current the anode is delivering. These control and monitoring systems have been used in experimental systems. In routine installations, none of this will be necessary. The amount of anode consumed can be assessed by occasional coring and the anode replaced when necessary.

A comparison of monitoring techniques and of the merits and limitations of impressed current cathodic protection systems is given at the end of the chapter.

6.2.3 The history of cathodic protection of steel in concrete

The principles of sacrificial anode cathodic protection were discovered by Sir Humphrey Davy in 1824. His results were used over the next century or so to protect the submerged metallic parts of ships from corrosion. In the early decades of the 20th century the technology was applied to underground pipelines. When it was found that the soil resistance was too high, impressed current cathodic protection was developed.

The problem of corrosion of steel in concrete was first ascribed to stray current flows from trams and DC railway systems (Hime, 1994). Once chloride, in the form of deicing salt, was identified as the major culprit (when trams disappeared but corrosion increased), an enterprising engineer in the California Department of Transportation (Caltrans) took a standard pipeline cathodic protection design and 'flattened it out' on a bridge deck.

The system was straightforward. One of the popular impressed current pipeline cathodic protection anodes of that time was made of a corrosion resistant silicon iron, surrounded by a carbon cokebreeze backfill. A well was dug near the pipeline, the anode put in surrounded by the backfill and the system connected to a DC power supply, with the negative terminal connected to the pipeline to make a cathodic protection system. Richard Stratfull took 'pancake' silicon iron anodes, fixed them on a bridge deck and applied a carbon cokebreeze asphalt overlay (Stratfull, 1974). The systems installed in 1973 and 1974 were reviewed in 1989 and were still working (Broomfield and Tinnea, 1992).

The author conducted some of the earliest research in the UK for the UK DoT on conductive coatings for cathodic protection (Geoghegan *et al.*, 1985). The first UK trial was designed and installed by the author on Melbury House above Marylebone Station in London for British Rail in 1984. This included remote control and monitoring by modem link (Broomfield *et al.*, 1987).

Four trial systems were designed and installed in 1987 for the UK DoT on the Midland Links motorway system around Birmingham, Britain's second largest city. Those systems are still running (in 1996) and about 100 cross-beams have been protected (Unwin and Hall, 1993) with a further 400 or more expected to be protected on the Midland Links before the year 2000. Early systems were also designed by the author in 1986–7 and were installed on bridges and buildings in Hong

Kong and marine structures in Australia. There is now a large number of systems on bridges and other structures in the Middle East.

By the early 1990s a major UK based installer of cathodic protection systems had installed over 20 000 m^2 on about 12 structures, mainly in the UK and the Middle East. In 1990–91 they installed about 11 000 m^2. A major UK consultant on cathodic protection for reinforced concrete has overseen the installation of about 64 000 m^2 on about 24 structures in the UK and the Far East. The largest anode manufacturer, Elgard Corporation, has supplied over 400 000 m^2 world wide including 60 car parks and 400 systems on bridges and tunnels (1993 figures).

Since those first systems were applied in the 1970s systems have been developed and applied to bridge decks, substructures and other elements, buildings, wharves, and every conceivable type of reinforced concrete structure suffering from corrosion of the reinforcing steel. Systems have also been applied to steel in mortar in stone clad structures.

Anodes have been developed in the form of conductive coatings, metals embedded in concrete overlays, conductive concrete overlays and probes drilled into the concrete. Anodes continue to be developed, applied in new configurations and to new structures. In the next section we will discuss the major components of the cathodic protection system, and particularly the anode systems that are available as these are the most prominent part of the cathodic protection system. Judicious choice of cathodic protection anode can maximize the cost effectiveness of the system.

The reason for choosing cathodic protection is almost always cost effectiveness. The cathodic protection system prevents corrosion across the whole of the protected area of the structure, unlike localized patch repairs. If repeated cycles of patch repairs are too expensive or unacceptable, then a properly maintained cathodic protection will usually work out as more cost effective in the long term. If the structure only has a short-term future and patch repairing is acceptable and inexpensive, then cathodic protection is not usually a suitable option.

6.3 THE COMPONENTS OF AN IMPRESSED CURRENT CATHODIC PROTECTION SYSTEM

The essential elements of a conventional impressed current cathodic protection system are discussed below.

The cathode is the the reinforcing steel to be protected. It must be continuous, i.e. electrically interconnected to allow current flow, and separated from the anode by an electrolyte (the pore water). If it is not electrically continuous then it must be made so by adding rebars, wiring or welding elements together. If there are 'shorts' to the anode

Impressed current cathodic protection systems

(usually where tie wires or shallow bars come into contact with the anode placed on the concrete surface or in holes or slots in the concrete) they must be removed, or the anode cut away. The electrical connection from the rebar to the cable leading to the power supply is usually made with a 'thermite weld' (Figure 6.5). This uses a mixture of carbon and aluminium oxide to produce a hot weld to bond the steel to the copper conductor.

Impressed current anodes have very slow or controlled consumption

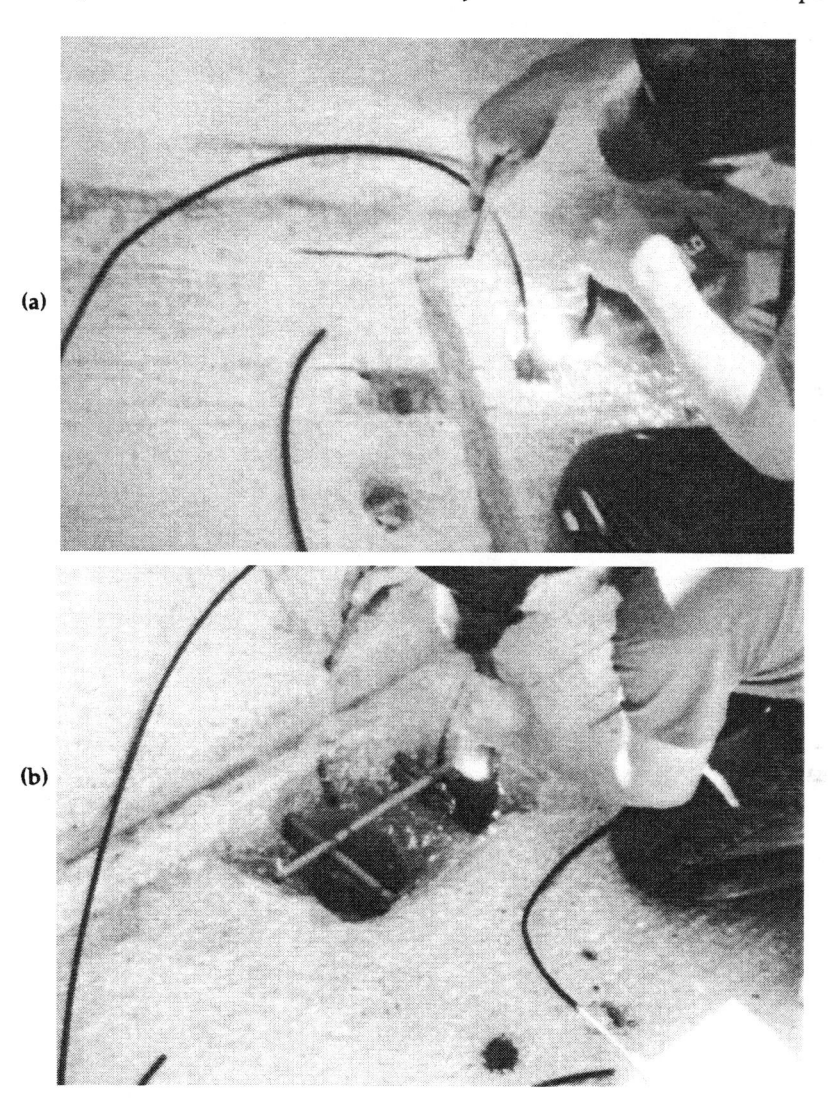

Figure 6.5(a) and (b) Thermite weld.

rates when the anodic reaction occurs on the anode surface. As the reaction consumes alkalinity (equation 6.5) and generates acid (equation 6.6) it can attack the anode and the concrete. Minimizing the level of current is therefore important in maintaining a good anode to concrete bond. Types of anode are described below. The function of the anode is to spread the current to all areas to be protected having converted the electrical current from the transformer/rectifier (DC power supply) to an ionic current that flows from the anode to the cathode so that the cathodic reaction will occur on the reinforcing steel surface, suppressing corrosion.

This is usually done with a 'distributed anode', such as a paint coating on the surface, an expanded metal mesh across the surface encased in a concrete overlay or strips of anode in slots across the surface. Alternatively, a series of anodes can be embedded in the concrete among the rebars.

As gases are evolved at the anode (equations 6.4 and 6.5) the distributed anodes such as coatings or encased anodes are usually gas permeable.

The electrolyte is the medium through which the ionic current flows from the anode to the cathode. It can be soil or sea water for pipelines or ships. In the case of atmospherically exposed reinforced concrete structures it is the concrete pore water. In the case of buried or submerged concrete it will be first water or soil and then the concrete pore water. Our discussion will concentrate on atmospherically exposed concrete structures, where concrete pore water is the electrolyte.

It is important to realize that only some of the pores will contain water and most of them will not be filled but probably lined with water. The tortuosity and small size of the pores in the concrete gives a higher resistivity for inland atmospherically exposed concrete structures than for marine or soil cathodic protection systems. Added to this is the fact that not all the pores are 100% full, or even lined with water. This makes a very unusual highly oxygenated, stagnant, high resistivity, alkaline medium. This contributes to the unique requirements of cathodic protection systems for atmospherically exposed reinforced concrete structures.

Problems can arise if the concrete around the anode dries out. Initially this gives rise to a requirement for a higher driving voltage (10 to 15 volts instead of the usual 1 to 5 volts). It is usually assumed that if the concrete dries out much more than this, then the concrete around the steel is too dry for corrosion to occur. This may not be true, but will not matter if the drying out is short term (say less than a month or so during a hot dry summer).

The transformer/rectifier (T/R or rectifier) is the DC power supply that transforms mains AC to a lower voltage and rectifies it to DC. The positive terminal is connected to the anode and the negative to the cathode. The level of the output is controlled as described below. T/Rs can be run at constant voltage, constant current or constant potential (against a half cell). They can be adjusted manually, automatically by circuitry or computer control or remotely using a telephone line and modem link or similar remote connection as described later.

Transformer/rectifiers for conventional cathodic protection systems can be very large and powerful, capable of delivering hundreds of amps, with oil cooled transformers. However, for steel in concrete our requirements are more modest. A graph of current requirement versus chloride content is discussed in Section 6.5. Most systems are designed for a current density of about 10 to 20 mA per square metre of steel surface, with allowances made for the reduced current flow to lower layers of steel. Allowances must also be made for the voltage drops down the connecting cables. It is usual to restrict the maximum voltage to a low level, usually around 12 to 24 V DC. This ensures that there is negligible risk to humans or animals of electric shock or harm from the current.

The problem with selecting a transformer/rectifier at the design stage is that it must be powerful enough to stop corrosion, but not so powerful that all adjustments are made in the first 10% or less of the output. It is common to overestimate the current requirement and then add a safety factor of 50%. If only say 10 amps output is needed and the T/R can deliver 50 amps, this can lead to difficulties in control, as fine adjustments are needed in the first 20% of the range. There will also be inefficiency in output, with electrical energy lost as heat. The electrical engineering of transformer/rectifiers is covered in some of the main texts and standards on cathodic protection and will not be discussed in detail here. In general terms, the systems for protection of steel in concrete are generally full wave rectifiers with smoothing to minimize interference and any possible adverse effects on the anode.

A continuously variable output is usually specified. Most cathodic protection systems are run under constant current control, or constant voltage. T/Rs should usually be able to run in both modes. Control by constant half cell potential against an embedded reference electrode is rarely used for steel in atmospherically exposed concrete.

Transformer rectifiers must be adequately protected from the elements. If they are sited outside, they are usually in steel enclosures (suitably corrosion protected!) with adequate ventilation and protection against vandalism or other damage. They often incorporate lightning arresters and other protection for the public and the inspector of the system, with suitable fusing and earth protection. Heating to avoid condensation is also specified in some environments.

Other sources of DC power are available in principle but are usually too expensive for use on large installations. Wind or water turbines, solar arrays and batteries have been proposed and experimented with, but few have been used in large scale practical applications.

Probes are half cells or other small probes embedded in the concrete. These measure the effect of the cathodic protection current and enable operators to set correct current and voltage levels for the system. They are discussed more fully in Section 6.5 on control criteria.

6.3.1 Cathodic protection anode systems

The anode is a critical part of the cathodic protection system. It is usually the most expensive item and has the highest cost of installation with most disruption to the operation of the structure (although far less than most alternative concrete repair systems). It is crucial that the correct anode is chosen and that it is applied properly.

Anodes can be divided roughly into two groups, with some crossover. The first group to be developed was for bridge decks with vehicles running over them. They are durable anodes with concrete or asphalt overlays. Some can also be applied to substructures. Substructure anodes can be applied to vertical and soffit surfaces. They may be less resistant to wear and abrasion than the deck anodes.

6.3.2 Deck anode systems

Deck anode systems are generally embedded in a concrete overlay or recessed into slots cut in the deck. There are advantages and disadvantages to both approaches. Overlays increase the thickness of the deck, which can increase the deadload and reduce clearances. Anodes recessed in slots avoid these problems but they are difficult to cut and must be close together (about 300 mm) for adequate current spread.

The first anodes, developed by R. Stratfull of Caltrans, were silicon iron primary anodes in contact with a conductive cokebreeze asphalt overlay (the secondary anode). This design was based upon cathodic protection designs for pipelines where a silicon iron anode is embedded in a carbon cokebreeze backfill to give a large contact area and low resistance. The anode is then linked, via the transformer/rectifier, to the pipeline to be protected. The Stratfull anode design is shown in Figure 6.6. A modified form of the conductive asphalt system is still used in Ontario, Canada. In some cases, carbon anodes have been used instead of silicon iron. The principal problem with this anode is that the cokebreeze 'secondary anode' is permeable and water can get trapped at the concrete–anode interface, leading to freeze–thaw damage. This means that the concrete must be checked to ensure that there was adequate air entrainment during construction. The anode also adds to the

Impressed current cathodic protection systems 121

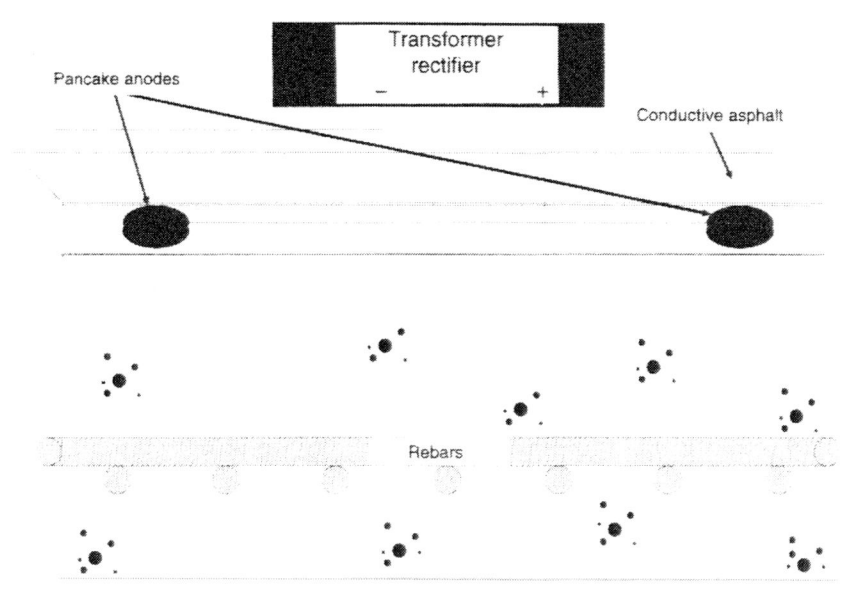

Figure 6.6 Conductive asphalt and pancake anodes for decks.

deadweight of the structure and can change the riding surface level on bare concrete decks.

The next development was a conductive polymer grout (carbon loaded resin) put into parallel slots cut into the surface (Figure 6.7). The aim of the slot system was to reduce the weight of the anode and to eliminate the change in deck height. However this system requires good concrete cover to the reinforcing steel to avoid short circuits between the anode and the reinforcing steel. The original polymer did not bond well to the concrete in all cases, and showed high consumption around patch repairs as shown (left hand side by barrier). The system has been superseded by coated titanium ribbon (Figure 6.8). The effort of cutting slots on 300 mm centres in the deck means that this anode is not used unless the deck level cannot be raised. This occurs under bridges and in buildings and other locations with fixed headroom requirements.

One of the first commercial, proprietary anodes for decks was a flexible cable with a conductive plastic round a copper conductor. The cable was 'woven' across the deck and then a concrete overlay applied. It was also used on substructures with a sprayed concrete overlay. This system was very popular in the 1980s but unfortunately the conductive plastic started to fail after about five years' service. This led to attack of the copper and, in some cases, expansion of the plastic which delami-

122   *Electrochemical repair techniques*

Figure 6.7 A carbon based slot anode system in Virginia, 1988.

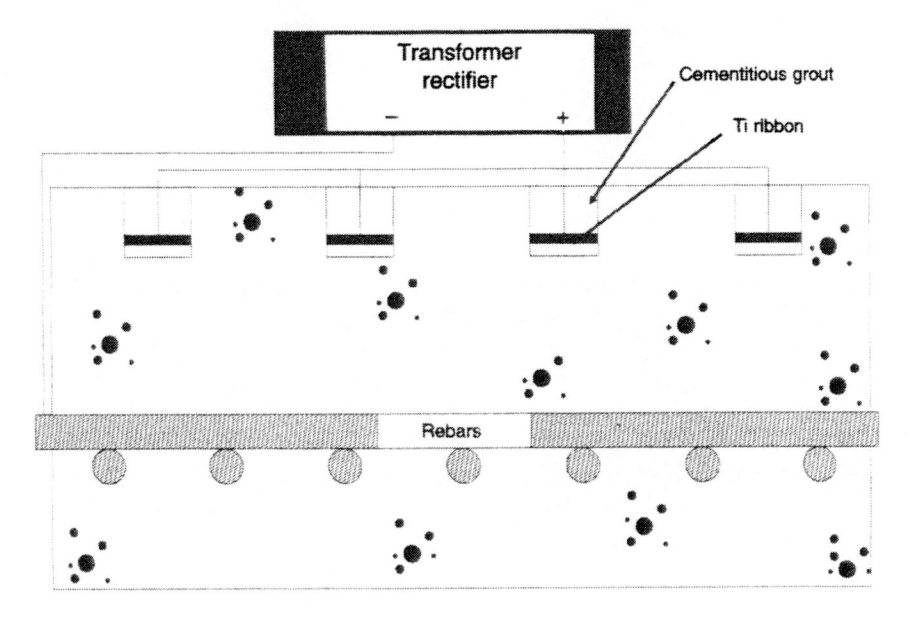

Figure 6.8 Slotted anode system.

nated and spalled the concrete overlay. This system is no longer available.

A fairly new anode system that has been used extensively in Italy is a conductive fibre concrete (de Peuter, 1993). This has the advantage of being a 'one component system'. It is applied directly to the concrete surface after preparation and repair of the old concrete. It is conductive and withstands modest levels of wear, abrasion and water action, but it requires an overlay if it is to receive vehicle traffic.

(a)

(b)

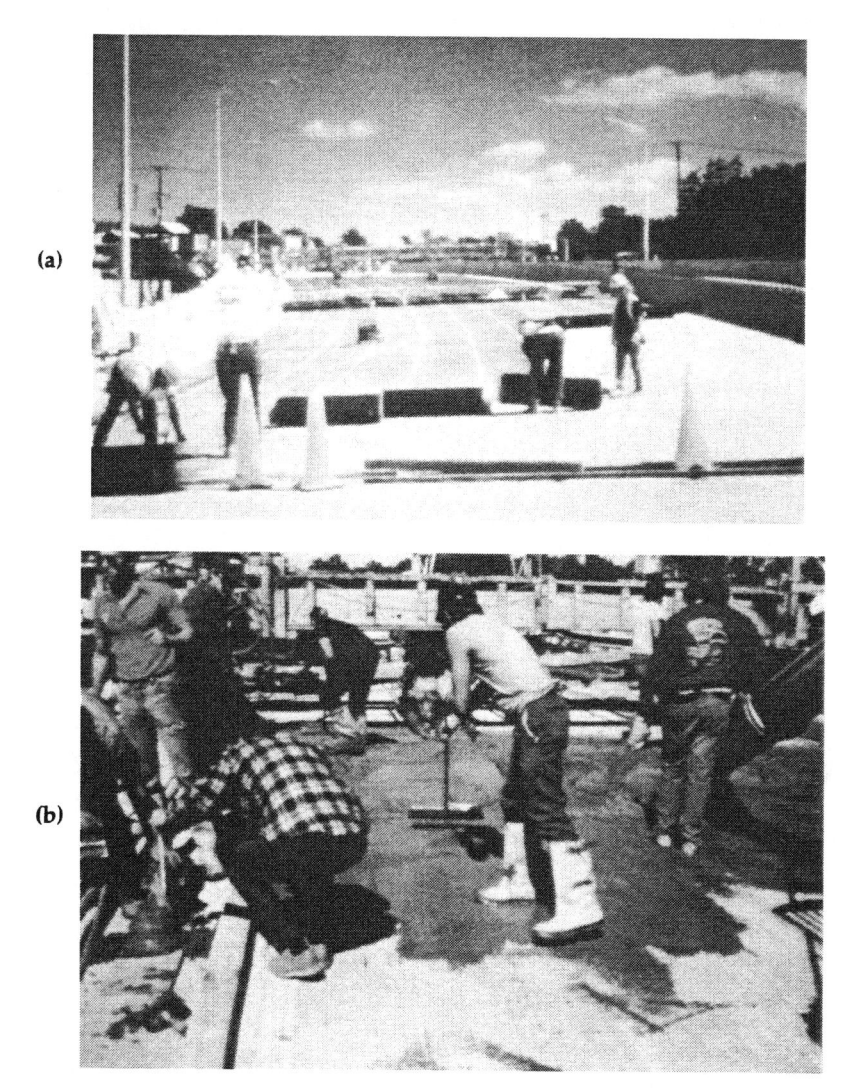

Figure 6.9(a) and **(b)** The Elgard™ titanium mesh anode system.

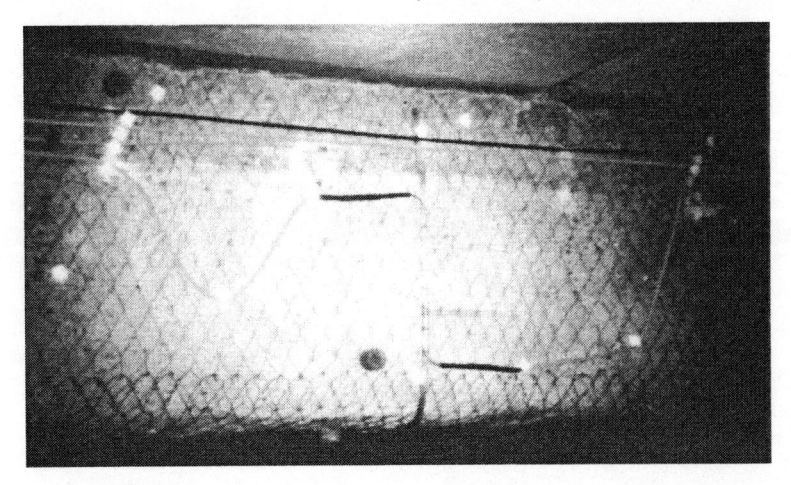

Figure 6.9(c) Elgard mesh on a substructure beam.

One of the most successful commercial anodes is expanded titanium mesh with an activated precious or mixed metal oxide coating. This also comes in the form of an expanded titanium mesh, strips and other configurations. It is fixed onto the surface, usually with plastic fixings and a cementitious overlay applied (Figure 6.9). The system can be used on decks and vertical or soffit surfaces. The anode itself is extremely durable but the application of the overlay is crucial to its durability. Figure 6.9(c) shows the anode awaiting a sprayed concrete overlay after being fixed to a beam. In Figure 6.9(d) the problems of overlay adhesion are seen where wave action has removed the overlay at the base of a column on a bridge in North Carolina.

6.3.3 Anodes for vertical and soffit surfaces

The expanded titanium mesh is one of the 'cross over' anodes. For soffit or vertical surfaces sprayed concrete is normally used to overlay the anode. As stated in Section 5.2 on patch repairs, sprayed concrete application requires considerable expertise to obtain a good bond between the overlay and the parent concrete. This is done by preparing the original surface (grit blasting scabbling or similar roughening and cleaning) and standing the mesh off the surface so that concrete rebound from the anode is minimized. The highest standards of 'on site' quality assurance and control are needed to get 100% adhesion of the overlay.

The mesh has also been mounted in permanent formwork and then grouted up by pumping in concrete or mortar from above or below. This avoids the quality control problems of sprayed concrete but gives

Figure 6.9(d) Storm damaged overlay on a titanium mesh system.

more engineering problems and can increase the cost of the permanent formwork. Electrical connection is made to the anode with strips of titanium welded to the mesh.

A platinized titanium rod probe anode is also available to provide local protection when embedded in a cokebreeze backfill (to minimize acid attack of the concrete) in a drilled hole in the concrete. Figures 6.10 and 6.11 illustrate the anode and its use. As this is not mounted on the surface there are two challenges to correct installation. The first is to ensure that there is no short circuit between the anode and the rebar. This is accomplished by using a special 'cover meter' which when placed in the drilled hole will warn if there is reinforcing steel too close to the hole. The second issue is screening of rebars. The anodes must be

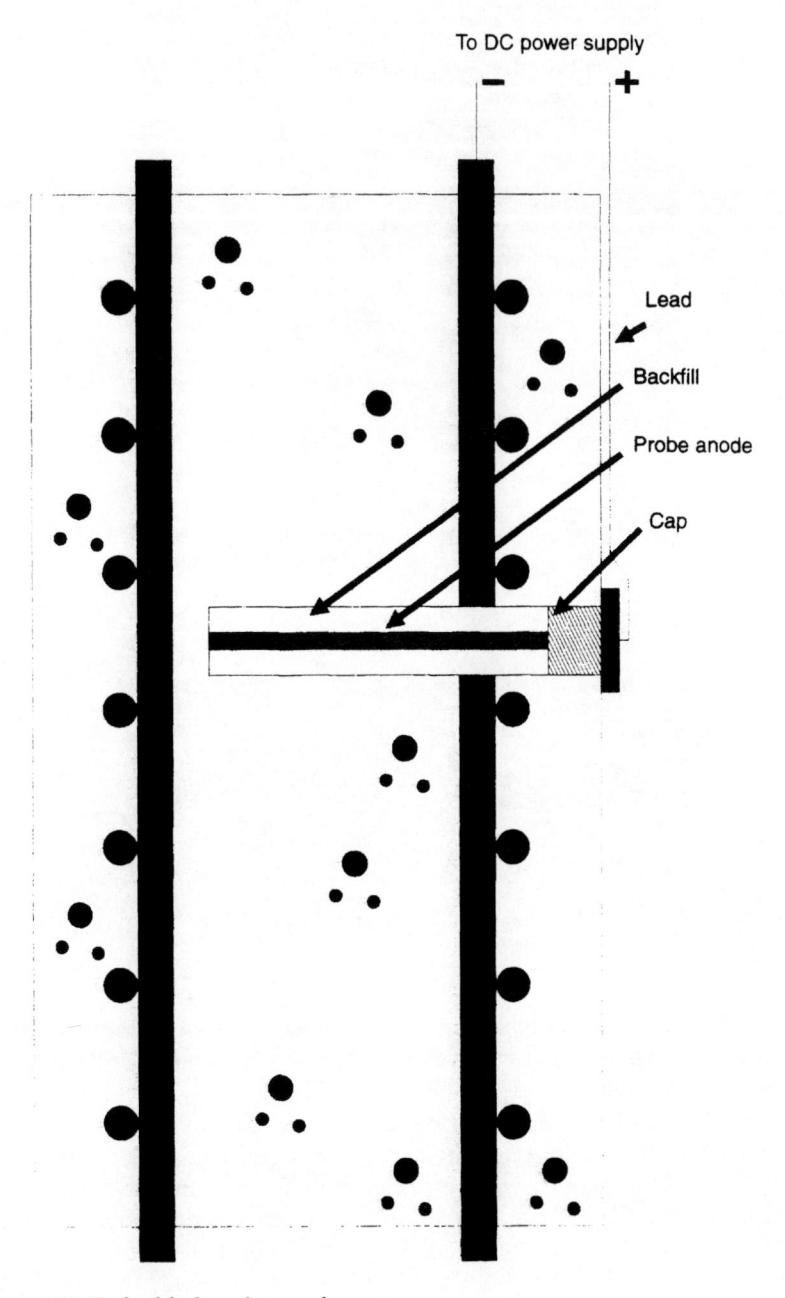

Figure 6.10 Embedded probe anode system.

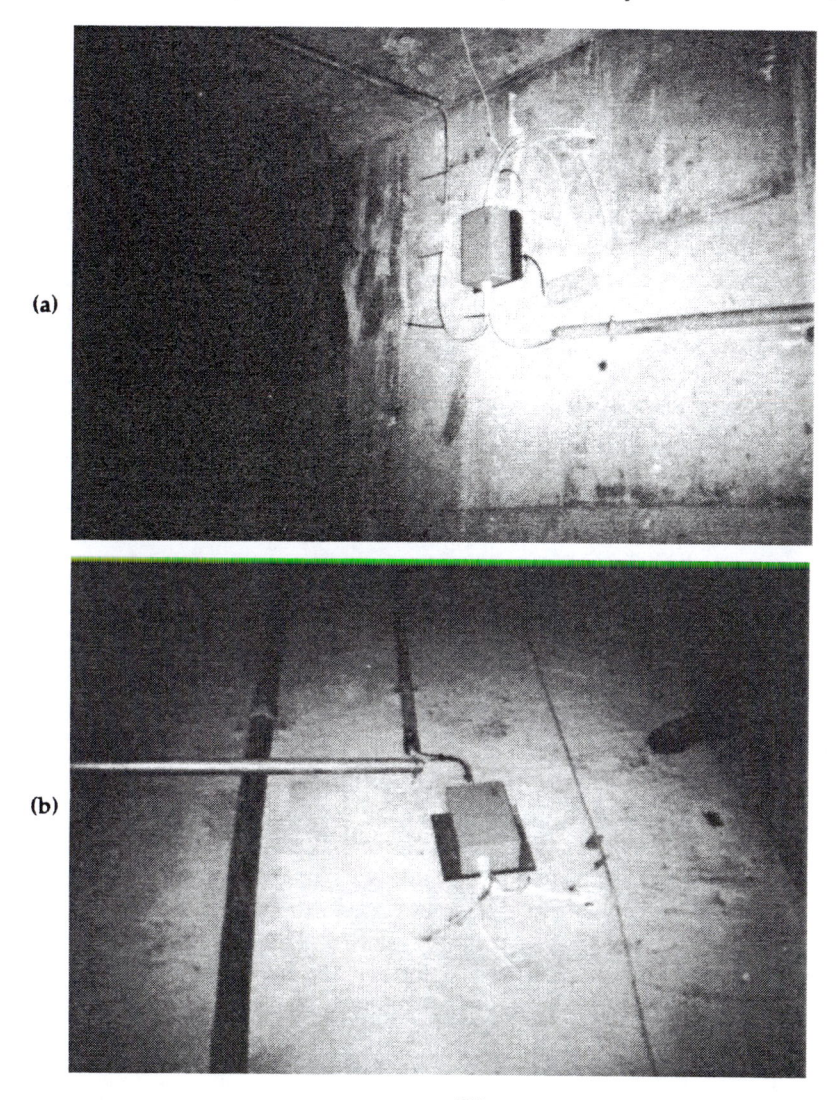

Figure 6.11(a) and (b) The DurAnode™ platinized titanium probe anode system being installed inside a hollow bridge in Norway, 1995.

distributed throughout the structure in such a way that there is adequate protection to all the corroding reinforcing steel.

This system can be installed in a deck to protect the steel under a trafficked surface. This is best done from below so that the system is not exposed to traffic impact and wear. If the steel is too congested then probes can be inserted above and cables run in slots in the deck surface.

6.3.4 Conductive coatings

The other type of anode to achieve considerable commercial success is the conductive coating. Coating anodes were developed from the carbon loaded paints used in electronics and specialist electrical applications and developed for space heating in the 1960s. Poor results for the application of these anode systems were reported in North America in the 1980s, with the first generation of coatings debonding and flaking off.

Research carried out in the UK for the DoT led to development of a system that was durable in UK conditions. The North American manufacturers developed their materials further and improvements in concrete surface preparation have led to more durable systems. A range of coatings is now available. Most of those with a proven UK track record are solvent based. Water-based materials are being developed in the UK and Europe and North America and are being used throughout the world.

Coatings do not have the durability of the titanium mesh with overlay. However, they are cheaper, easier to apply and can be repaired and maintained easily. They are not suitable for continuously wetted, abraded or trafficked surfaces. They require excellent surface preparation to get good adhesion. They are usually applied by brush, roller or airless spray.

The paint systems require an electrical connection. Given their modest conductivity this usually consist of a series of wires (platinized titanium, platinized niobium copper, carbon fibres or other comparatively inert materials) running through the coating at a separation of 0.5 to 1.0 m. Unfortunately the nomenclature used for the early anode systems has been misapplied to coatings. In the conductive asphalt system and the pipeline system it is based upon, the silicon iron (or carbon) anode is called the primary anode. The cokebreeze material (backfill or asphalt) is called the secondary anode. Likewise the connection wire in the paint system is often referred to as the 'primary anode' and the coating as the 'secondary anode'. This considerably inflates the significance of the connection wire and diminishes the importance of the coating which is really the anode. Figures 6.12 and 6.13 show the conductive coating anodes as applied to vertical surfaces.

There are several hundred bridge substructures protected with coating anodes in the UK and Europe, and dozens of buildings. There are also hundreds of parking structures in the USA and Canada with conductive paint coating anodes applied to them.

6.3.5 Thermal sprayed zinc

Given the early problems with conductive coatings in North America, Caltrans looked for an alternative. They experimented with electric arc

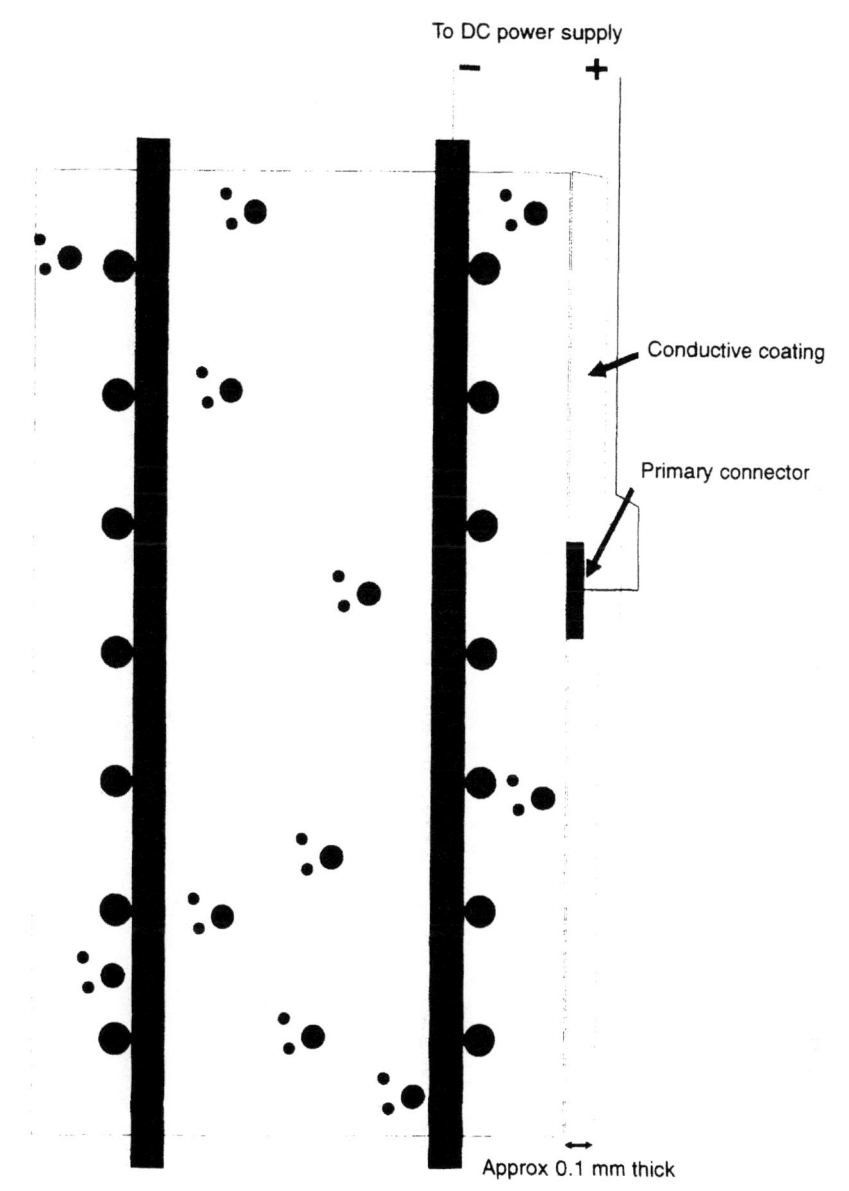

Figure 6.12 Conductive coating anode system.

and flame sprayed metals and found zinc to be most effective. Many arc sprayed zinc systems have been applied in the USA. They are particularly popular in marine substructure conditions where it is difficult to apply other anodes. Florida Department of Transportation has applied several such systems and Oregon DoT have sprayed two very large

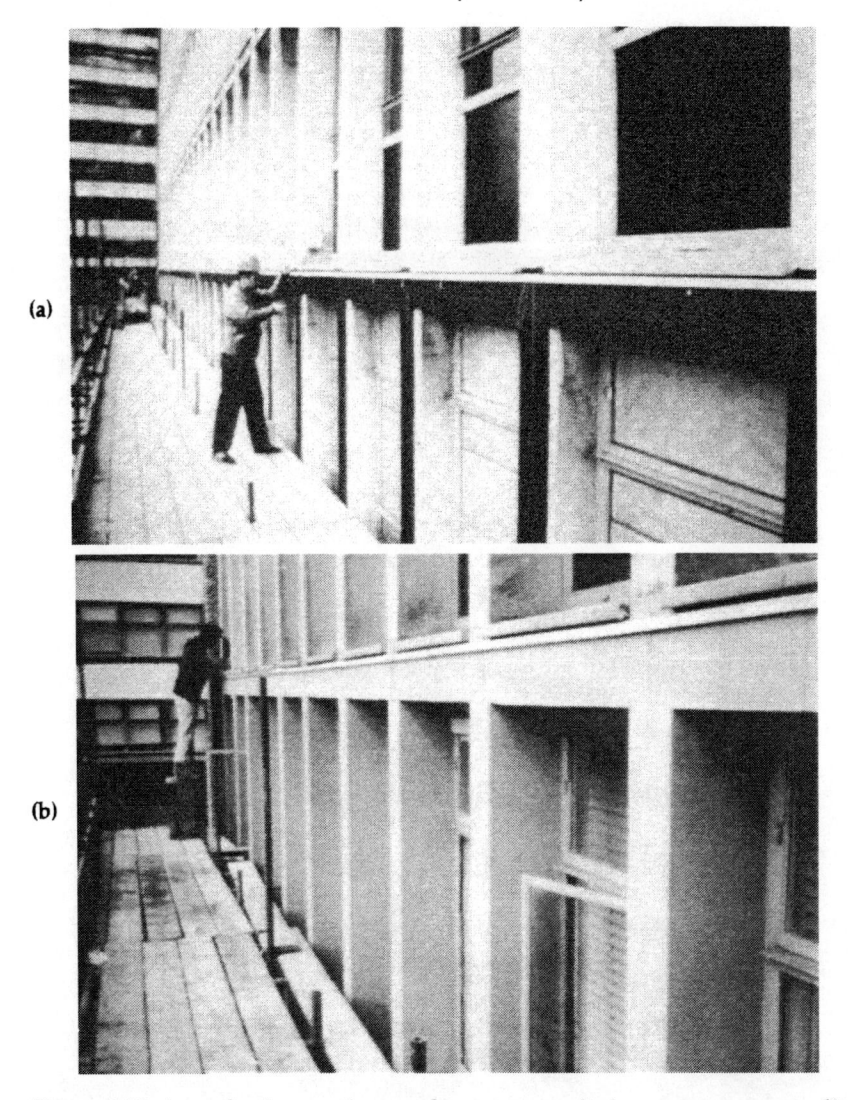

Figure 6.13 A conductive coating anode system applied to a government office building in the UK, (a) before and (b) after cosmetic top coat application.

bridge substructures over the past two years. The Yaquina Bay bridge has a 26 500 m^2 system applied and Cape Creek is about 13 000 m^2. The electric arc sprayed system is preferred over arc spraying in North America due to its rapid deposition rates (Figures 6.14, 6.15).

Sprayed zinc has been experimented with in the UK, generally using the simpler flame spray system, but there is some concern about the rise in resistance seen on some systems. This may be due to a build up

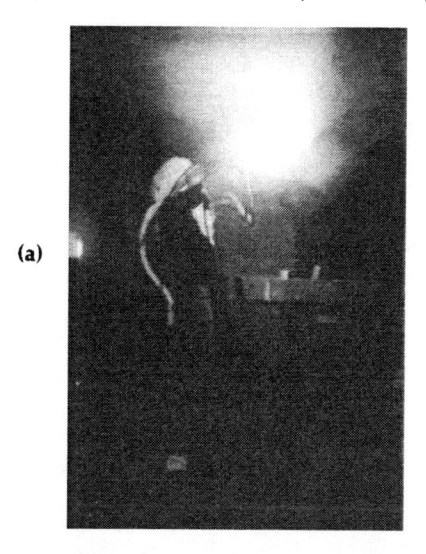

Figure 6.14 (a) Thermal sprayed zinc anode being applied using an electric arc. **(b)** The bridge substructure can be fully enclosed to avoid environmental contamination.

of corrosion products between the zinc and the concrete, or to treatment of the zinc after application to protect it from atmospheric corrosion. Zinc, of course, is not inert and is consumed by corrosion from the atmosphere and chlorides. The anodic reaction also consumes the zinc and gives rise to the formation of oxides and sulphates at the anode–concrete interface which may increase the electrical resistance between anode and cathode.

132   *Electrochemical repair techniques*

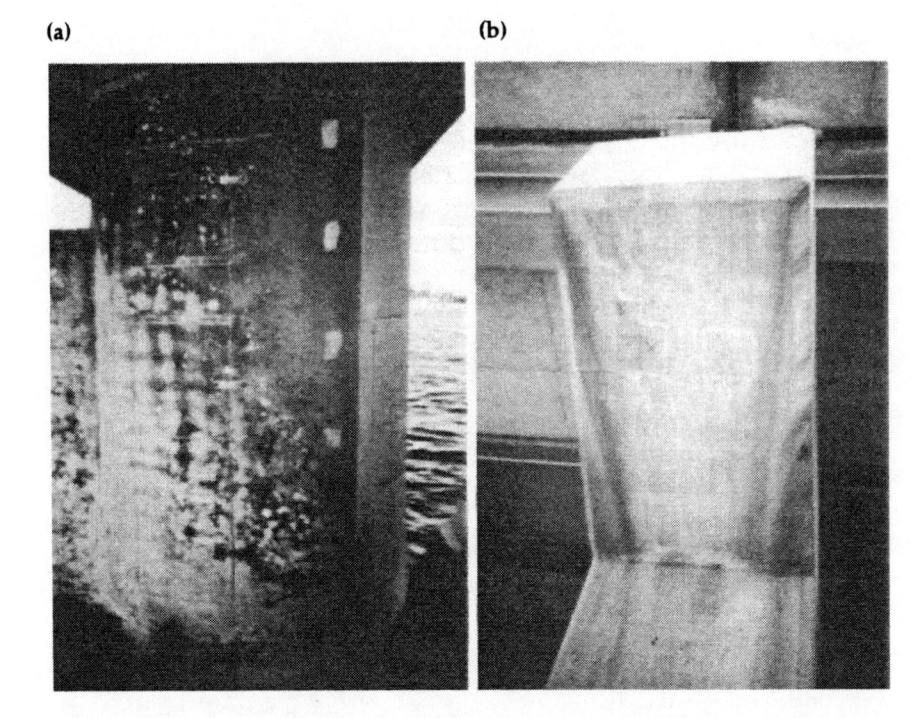

(a)   (b)

Figure 6.15 Thermal sprayed zinc systems for sacrificial CP on (a) a Florida Keys bridge column and (b) on a cantilevered beam showing cut out areas of short circuits between the steel and the surface due to tie wires etc. Florida DoT, North Dakota DoT, Corrpro and FHWA.

The use of high temperature spraying gives a very porous, open coating made up of small droplets of zinc metal. The coating has excellent adhesion to the concrete. The zinc coating looks similar to concrete and there is no need for extra protective or cosmetic coatings. The system is well established in the USA and trial sacrificial anode systems were described above.

Among the highway agencies in North America there seems to be a preference for arc sprayed zinc systems for protecting bridge substructures. In Europe the preference is for conductive paints. This may be because in the UK at least, guttering has been applied below leaking joints to minimize the amount of water running over the coatings and causing deterioration. This has extended their life. In North America this is often not done so the paints do not last as well as the zinc. Also the zinc can be applied to damp concrete surfaces, making it suitable for marine applications, and where the water continues to run over the concrete surface and degrade paint coatings.

As zinc coatings are highly conductive, electrical connection is made via a metal plate fixed to the surface and the zinc is sprayed over it. Very few connections are needed, but at least two should be applied in each zone.

One problem with zinc is its toxicity. In applications in Oregon, where there have been several very large applications, complete containment was required to prevent zinc dust from entering the atmosphere or contaminating the ground water. Applications in Florida have not required such stringent controls, reducing the costs considerably.

Other metals can be sprayed as well as zinc. At the time of writing the first trials are in place in Oregon using an arc sprayed titanium anode which is then activated once applied by spraying a proprietary chemical solution on to the surface (Bennett et al., 1995). This makes an anode similar (at least in performance) to the titanium mesh anode. This could be more durable than zinc and comparably priced as a lower volume of material is required. There do not seem to be any environmental restrictions with either the sprayed titanium or the activation solution.

6.3.6 Clamp on systems

One of the earliest experiments with a clamp on anode system was on Oregon Inlet bridge in North Carolina. This system consisted of plastic strips with a conducting gel on the inside that were fixed with plastic screws on to the bridge substructure. The problem was the even flow of current as there were too few anodes, too widely spaced apart (Figure 6.16). This system was not pursued and the bridge was destroyed in a gale around 1990.

Severe corrosion of reinforced concrete bridge substructures in the splash zone in Florida coastal conditions has led to work on systems for use in those conditions only by Florida DoT. Several clamp on systems have been on trial including conductive rubber mats with zinc metal and wood or recycled plastic clamps with zinc or titanium mesh. The sea water ensures a good electrical (ionic) connection to the concrete. The systems work from low tide level, or below, up to the top of the splash zone. Some have been in operation for several years. A system using tape and plates to secure the titanium mesh anode has also been experimented with in Australia, along with a screw on plate (Figure 6.17).

6.3.7 Sacrificial anode systems

Sacrificial anode cathodic protection works on the principle that different metals have different electrochemical potentials. Once more we

Figure 6.16 The experimental clamp on anode system on a bridge substructure.

return to the simple electrical cell where an electrical potential is generated by having two different metals in ionic contact in the cell.

Zinc in acid conditions has a potential of about −1.1 V against a copper/copper sulphate cell. As passive steel has a potential of about −100 mV and active steel is usually −350 to −500 mV, a zinc anode has 0.5 to 1.0 V electromotive force (emf) to cathodically protect the steel, assuming that the zinc is kept active. The potential of the zinc will change in the same way as the steel does, with the corrosiveness of the environment. Alloys are used in some cases to reduce the formation of oxides that will increase the electrical resistance between the anode and the steel. This issue of 'passivation' of the anode is a problem with anodes in alkaline environments such as concrete.

Aluminium and magnesium and their alloys are also used in sacrificial anode cathodic protection systems. One advantage of these alloys is

Figure 6.17 Experimental clamp on anode system trial on a section of a wharf substructure.

that they are lighter than zinc. However, their oxides and other corrosion products are voluminous and could also attack the concrete.

The two zinc systems described above, the thermal spray and the clamp on system, have also been tested as SACP for bridge substructures in marine conditions.

The earliest SACP systems were installed in the USA on bridge decks. The first was in 1978 using aluminium, and from 1976 to 1980 a study was done on a zinc system in Illinois. Both were metal embedded under overlays on bridge decks. The latter system gave 10–30 mA m^{-2} current and, after several years' operation, achieved 100 mV decay (see criteria below). Despite initial negative reports, its long-term performance appears to have been good with no maintenance throughout its life. It was removed in about 1990 when the deck was replaced.

Florida DoT decided to try the system on substructures, both on ordinary reinforcing steel in concrete and epoxy coated reinforcing steel (also discussed in Section 6.8). They have had encouraging results. Some work in Florida was sponsored under the SHRP program and the author is currently acting as a consultant on a major US Federal Highway Authority (FHWA) contract to investigate the feasibility of SACP in non-marine conditions (Whiting et al., 1995). Other researchers are continuing to explore the marine application of the thermal sprayed zinc system. A third contractor has developed a 'glue on anode' that could be used as an impressed current or sacrificial anode.

The limiting factors for SACP are the electrical resistance of the concrete and the performance of the anode. Zinc will produce about 1 V electromotive force (emf) when coupled to steel. If the resistance of the concrete and the anode to concrete interface can be kept low this will generate sufficient current to protect the steel. If anode corrosion products, drying out of the concrete or chemical changes increase resistance or decrease the emf, the steel will no longer be protected. The more active the initial corrosion potential of the steel, the smaller the driving voltage.

The advantages of SACP are that it does not need mains power, a T/R, instrumentation or much maintenance. Its shortcomings are the lack of control of the system, consumption of the anode and the need to keep resistance low. If present trials and research show the system to be feasible, this could encourage more engineers to try out this comparatively simple form of cathodic protection.

6.4 CATHODIC PROTECTION SYSTEM DESIGN

The following discussion is not intended to explain how to design a cathodic protection system from scratch. We will consider some of the major processes that a qualified corrosion engineer will go through in designing a system, showing how decisions are arrived at about the components and the system performance. Design should always be undertaken by a suitably qualified and experienced expert.

6.4.1 Choosing the anode

If cathodic protection is the chosen rehabilitation methodology then the correct choice of anode is vital. For non-marine applications the usual choice will be an impressed current system. If it is a wearing surface then coatings are usually excluded. This leads to the use of the conductive concrete anode or one of the titanium configurations in an overlay or titanium ribbons in slots.

The ribbon in slots is used on decks where a change in deck height or increase in load is not acceptable and there is sufficient concrete cover. Cutting slots at 300 mm intervals is expensive and difficult and so this system is only used if it is strictly necessary due to the constraints of clearance or load.

The main restrictions on the overlay type systems are caused by the cementitious overlay. The drawbacks of these anodes are:

- they increase the dead load on the structure (the concrete overlay);
- they change the profile and can reduce clearances (again the overlay);

- the overlay can be difficult to apply in restricted areas or with complicated geometries, especially when applied as a sprayed concrete overlay.

It is essential that good quality control is maintained to get a good quality, adherent overlay.

For mainly dry bridge substructures, buildings and other soffit or vertical applications coatings are often used. These have a lower life than the titanium based systems but are easily maintained and are cheaper. They are often more cosmetically attractive for buildings as masonry paint type overcoats are supplied. Thermal sprayed zinc can be used instead of the paint type coating. As the metal is sprayed hot, it will generally dry out a damp surface enabling it to bond, unlike the paint-type coatings. It is therefore preferred in marine or regularly wetted environments.

The advantages of the coatings are:

- negligible increase in dead load;
- can be applied to any geometry;
- choice of decorative finishes;
- cheap and simple to repair or replace.

Their restrictions are:

- limited resistance to wear and moisture;
- limited lifetime (about 10 years in most applications).

Sacrificial anodes can be used in continuously wetted environments. These systems are not yet fully understood but they seem to be successful as arc sprayed zinc on marine bridge substructures. They need less maintenance than the impressed current systems. One problem with the zinc system is the environmental impact during spraying. If enclosure is required during spraying, the costs can be very high.

Platinized titanium rod probe anodes can be used to apply current locally to inaccessible areas in conjunction with other anodes or they can be drilled into a series of holes running up a column or similar structure. Their main problems are ensuring that the reinforcing steel does not shield the current, making sure that the anodes do not contact the reinforcing steel or overprotect it locally, and ensuring that the wiring to series of anodes is not too complicated.

Table 6.1 gives a comparison of different anodes, their advantages and limitations.

Table 6.1 Types of anode

Types	Deck/subs/splash	Weight	Durability (years)	Comments
Ti mesh in overlay	All	Increase	20+	Durable, established system. Main problem is overlay application
Asphalt	Deck	Increase	20+	Durable and established. Freeze-thaw risk. Oldest systems in North America
Ti ribbon in slots	Deck	No change	20+	Need 40 mm cover. New but should be durable
Conductive concrete	All	Increase	10–20	New. Needs wearing course for traffic
Grout in slot	Deck	No change	10?	Poor durability. Needs 40 mm cover
Paint coating	Substructure	No change	10+	Avoid wear or wetting. Easily replaced or repaired
Sprayed zinc	Substructure	No change	10?	Can be used sacrificially. Will stand wetting. Environmental risk during application
Sprayed Ti	Substructure	No change	20+?	Very new. Lower environmental impact than Zn. More durable?
Clamp on	Substructure	Increase	10?	Very new. Several configurations
Embedded rods	Substructure or soffit	No change	10?	New. Need careful design and installation. Wire to each probe

6.4.2 Transformer/rectifiers and control systems

This is another vital part of an impressed current system. The T/R must be rugged and reliable with minimal maintenance requirements. It should be easy to maintain with good instruction manuals, circuit diagrams for maintenance and easy access to fuses and other replaceable components. Compared with pipeline or marine cathodic protection applications (steel piles, etc.) the power demand is modest. Steel in concrete needs less than 20 mA m^{-2} to provide protection, usually at less than 10 V. The power for a 100 W light bulb will protect 5000 m^2. This means that a single-phase, air cooled T/R will usually protect even the largest structure and power consumption is rarely an economic concern.

The capacity of the transformer/rectifier and of the anode can be estimated from Figure 6.18 taken from Bennett and Turk (1994). One possible use of this chart, if it can be sufficiently validated, may be to eliminate the need for probes and control systems for cathodic protection. Once the highest chloride level at rebar depth is found during the condition survey, then the current is set at the appropriate level and left. The only maintenance need would be to ensure that the system is working and the current is being applied fairly uniformly to the structure.

There are two opposing directions of T/R design at the moment. The first is to make a simple, rugged reliable design with high quality

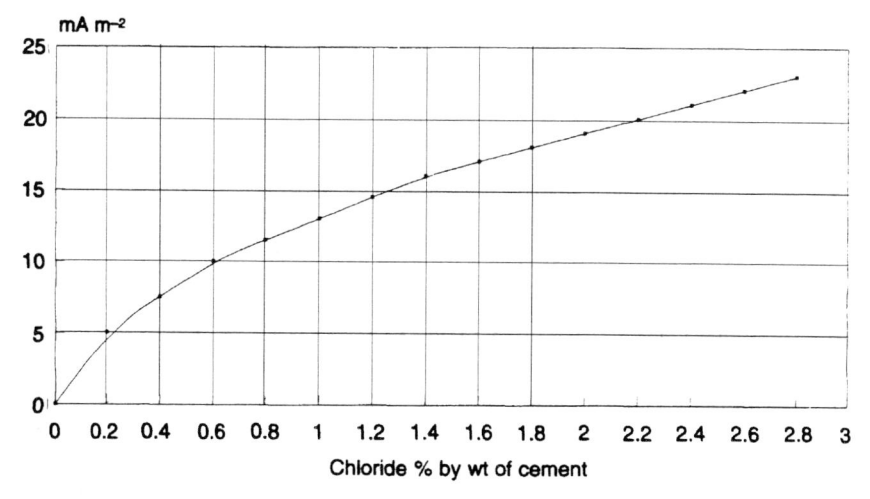

Figure 6.18 Cathodic protection current demand vs. chloride content at rebar depth.

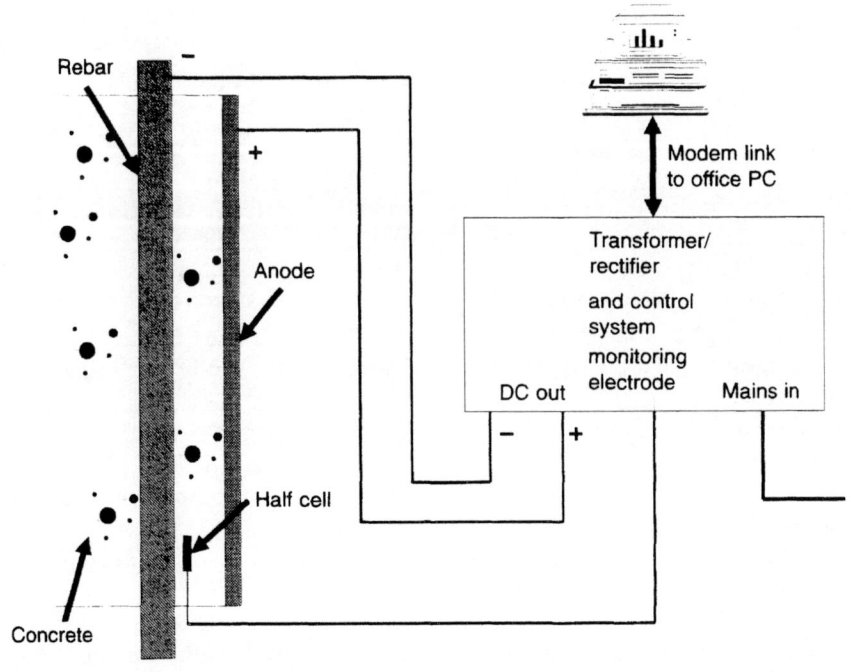

Figure 6.19 Schematic of cathodic protection for steel in concrete including modem link and remote monitoring.

components. This is checked manually every month or two and an annual 'service' carried out. The other is to attach a microprocessor that can monitor and control the system (Figure 6.19). This means that data can be collected remotely via a modem link and in some designs the system can be adjusted without such regular site visits.

The number of systems an organization has in operation is one factor in choosing remote control. It becomes more cost effective to collect and review data without site visits as the number of cathodically protected structures increases. The sophistication of the client and his consultant is another factor.

The reliability of the microprocessor system has not been reported in the technical press, although the author's first few remote control systems are still working satisfactorily after about 10 years. However, most of those are installed inside buildings in very benign environments.

Some systems are comparatively simple and will only monitor on and off potentials, current and voltage. It is not possible to change the current or voltage settings on some of these systems. With modern microprocessor technology, current remote control systems should be capable of monitoring on and instant off potentials, transformer rectifier functions, and conduct potential decay measurements (see 'control criteria' below).

6.4.3 Monitoring probes

In order to measure the effect of the cathodic protection current on the reinforcement, probes are embedded in the concrete. There are usually between one and five half cells per 'zone' or separately powered anode area. Ideally, half cells are located near the most actively corroding steel, and embedded in a way that does not disturb the concrete around that corroding steel. It has been North American practice to use 'salty' concrete to embed half cells. This in not done in Europe.

The most common probe is the embeddable half cell. A number of formulations are used but the most popular is the silver/silver chloride cell (Figure 6.20). Mercury/mercury oxide, lead/lead oxide and other formulations are also commercially available as well as carbon, coated titanium and lead. These last three are 'relative' rather than 'absolute' references. This means that they may stay stable over a few hours for a potential decay measurement, but the absolute values of the potential measured cannot be relied upon.

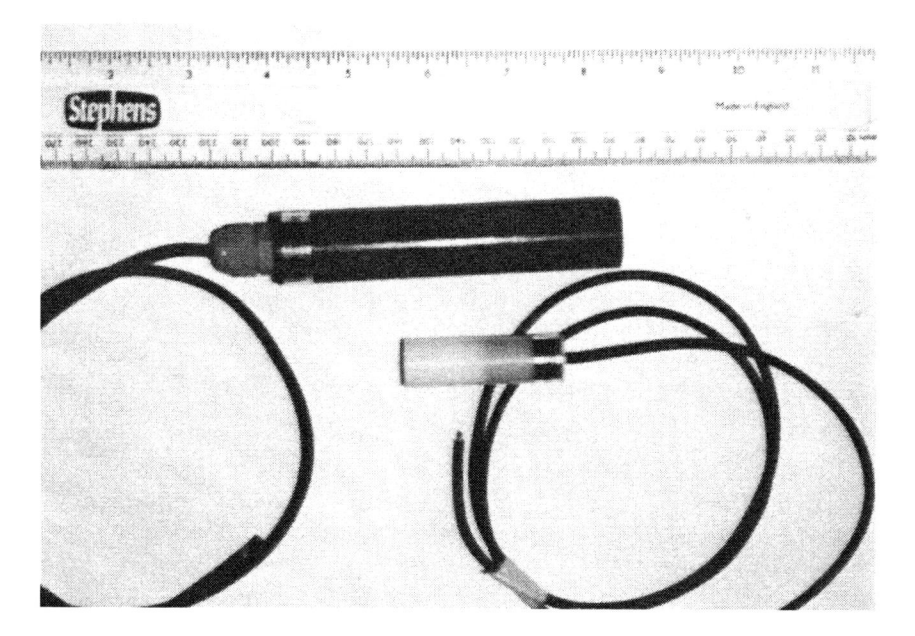

Figure 6.20 Embeddable silver/silver chloride half cells.

There has been a considerable discussion about reliability of electrodes in the North American technical press, but fewer problems have been reported in the UK and Europe. This may be due to extensive use in bridge and car park decks in America where traffic, water ponding, etc. reduce the half cell lifetime compared to vertical and soffit applications. More freezing and thawing at depth may also be a problem, or it may be the different designs available in different countries.

Due to the low reliability of some half cells in North America alternative probes are frequently used. These include the current pick up probe. A section of steel is embedded in the concrete. This picks up a proportion of the current and is used to gauge the effectiveness of the system.

The steel probe is sometimes embedded in an excessively salty patch. With no current applied, a macrocell current flows from the probe to the reinforcement if they are connected with an ammeter between them. As cathodic protection current is applied, the current reduces and then reverses. This is called the macrocell probe approach and is used to show that a very anodic area has been made cathodic. It is therefore assumed that the rest of the steel is cathodic, too. However this is

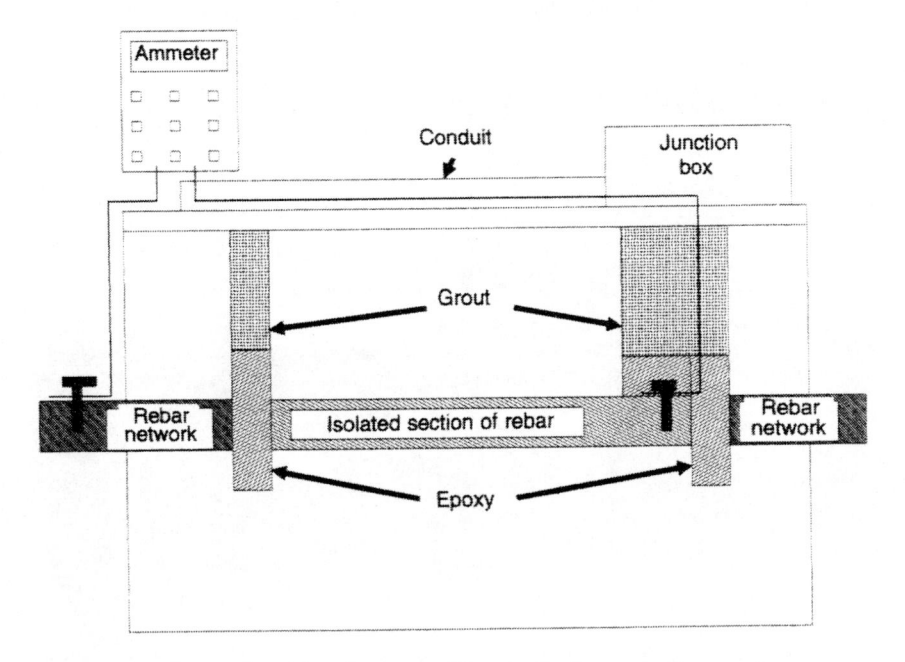

Figure 6.21 The null probe for controlling cathode protection systems constructed at the most anodic locations.

dependent upon the amount of salt added to the patch, and if the salt diffuses away then it may no longer be the most anodic area after a few years.

An alternative is to identify the most anodic area of the zone and to isolate a short section of steel to form a macrocell or 'null' probe without disturbing the concrete around the probe (Figure 6.21). This is more realistic than embedding a probe in salty concrete but suffers the same problem with chloride movement with time.

6.4.4 Zone design

One frequently asked question concerning cathodic protection systems is 'what happens where the anode finishes? Is there a risk of accelerated corrosion?' This is a valid question and the risk is supported by the Pourbaix diagram which shows areas of imperfect passivity, pitting and corrosion around the immune and passive regions (Figure 6.2). However, the author knows of no structure that is totally protected by cathodic protection. Most have anodes that end before the concrete does. No cases of accelerated corrosion have been reported between zones or at the end of zones.

It is rare for a cathodic protection system to consist of one continuous anode passing current from a single power supply. It is normal to divide the structure or elements to be protected into zones that are powered and controlled separately, and electrically separated by a gap of about 25 mm.

Zone design can be dictated by requirements for different anode types (coatings on walls, mesh on floors, etc.) They can be based on different current requirements, with high chloride areas divided into smaller zones to give each zone a fairly uniform current demand. They can be designed around different steel densities in different parts of beams or columns. In the USA, cathodic protection designs on decks have had zones in the 500 to 1000 m^2 range, and bridge substructures 100 to 500 m^2 (AASHTO, 1993). This probably has more to do with geometry of deck slabs and substructures than with variations in local current demand.

6.5 CONTROL CRITERIA

We have stated that the transformer/rectifier must deliver 10 to 20 mA m^{-2}. We have also seen that we must minimize the anodic reaction to reduce acid attack at the anode/concrete interface. This can reduce bond and thus increase the resistance of the circuit. It is therefore essential that we apply enough current to stop corrosion but only just enough. The control criteria for cathodic protection of atmospherically exposed concrete have been the subject of considerable debate.

At the beginning of Section 6.2 on cathodic protection, it was explained that cathodic protection works by injecting so many electrons into the reinforcing network that the anodic reaction cannot occur so corrosion stops. How can we show that this has happened? In cathodic protection of steel in soil or water it is usual to adjust the current to achieve a potential of −770 mV or −850 mV against a copper/copper sulphate half cell on the surface as the system is switched off (the instant off potential). However, these criteria are not appropriate for steel in concrete for a number of theoretical and practical reasons. Two of the practical reasons are the difficulty in accurately measuring an absolute potential over a number of years when reference electrode calibration may drift, and the fact that if an absolute minimum (or maximum negative) potential is achieved throughout the structure then some parts of the structure will be overprotected as the corrosion environment varies so rapidly and severely across a high resistance electrolyte like concrete.

It was also stated that a theoretical definition of effective cathodic protection is to depress the potential of the most cathodic areas below that of the most anodic areas. However, as time passes and negatively charged chloride ions move away from the negatively charged steel, the most anodic areas will move. Variations in the resistance of the concrete will mean that current flow from the anode is more likely to reach the anodic areas of steel and polarize them, rather then the passive, cathodic, areas (especially if they are new patch repairs with higher resistance concrete).

In practice, the best control criterion is based on a potential shift. Theory and experiment tell us that a shift in potential of 100 to 150 mV will reduce the corrosion rate by at least an order of magnitude (Bartholomew *et al.*, 1993; Bennett and Turk, 1994). This is based on the linear polarization theory discussed in Section 4.11 on corrosion rates. Field evaluations have shown that this stops all further signs of corrosion damage in cathodically protected structures.

This has lead to a number of criteria because of the practicalities of measuring potential shifts. The potential shift can be measured when switching on the system and switching it off. It must be measured as an 'instant off' potential when the system is on so that the cathodic protection current, or 'iR drop' as it is called, does not interfere (Figure 6.22). This means that the half cell potential, without the effect of the cathodic protection current, is measured between 0.1 and 1 s after switching off the current.

In the laboratory, the iR effect disappears within a millisecond or less. However, on a large structure with effectively a colossal leaky capacitor

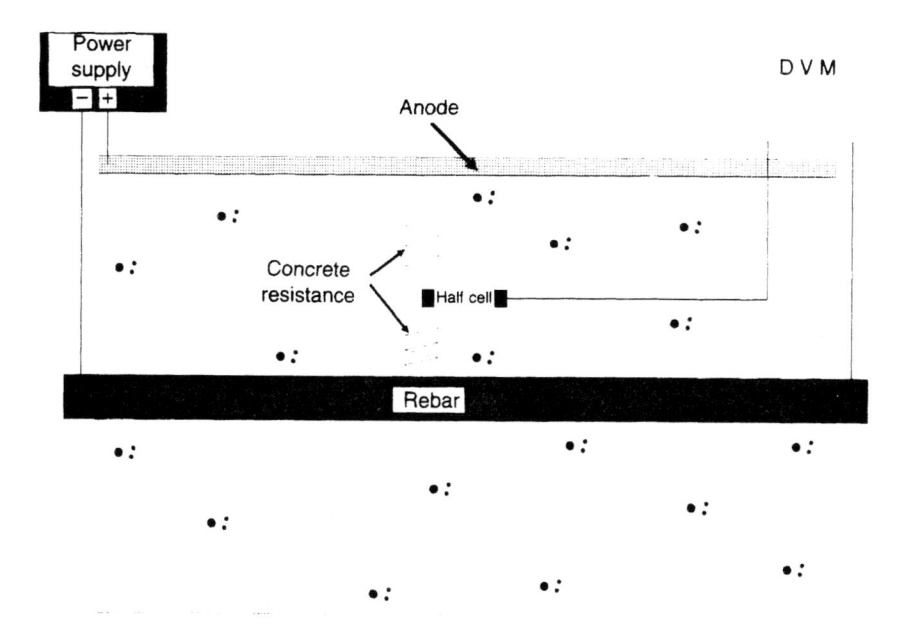

Figure 6.22 The iR drop phenomenon. While CP current is on half cell reads $V + iR$ where iR is resistance contribution from the concrete, depending upon exact location of half cell between anode and rebar.

made up by the anode and the cathode, the effects take far longer to decay, so usually 0.3 to 0.5 seconds is the delay. As a digital voltmeter samples three times a second, the first or second reading after switch off is used. For automatic logging systems it is advisable to use a site oscilloscope to check the shape of the curve on all cells and select a suitable interval during the commissioning of the system.

The steel in concrete system polarizes and depolarizes very slowly compared with many other cathodic protection systems. However, the potential is susceptible to changes in the environment (temperature, humidity), as well as the applied current, so readings must be taken within a time period while the environment around the structure is reasonably stable.

Therefore, the 100 to 150 mV shift in potential is often measured from instant off to a period typically four hours later. This is measured at an actively corroding (anodic) location. Measurements are made throughout the four-hour period so that the depolarization curve can be plotted. If depolarization continues at four hours then it is reasonable to expect that the 100 to 150 mV criterion has been achieved.

A number of things happen when cathodic protection is applied to

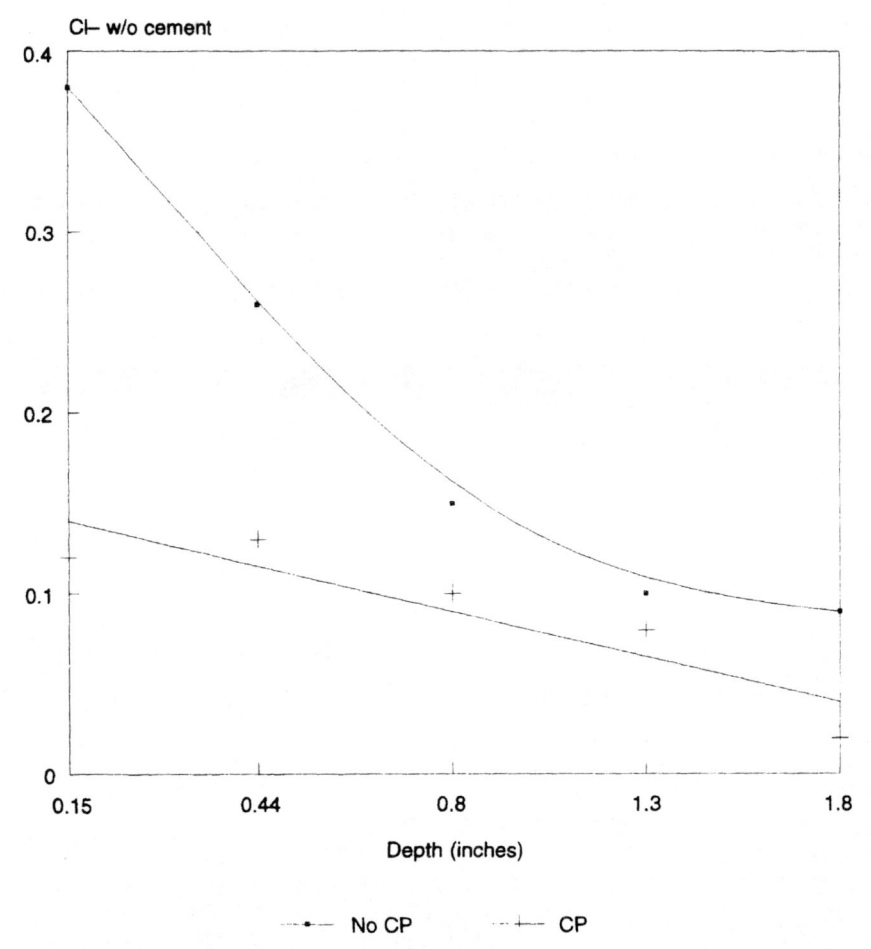

Figure 6.23 Comparison of chloride profiles in CP and no CP areas: Yaquina Bay bridge soffit (Broomfield and Tinnea, 1992).

steel in concrete. The negatively charged chloride ions move away from the negatively charged reinforcing steel (Figure 6.23) and hydroxyl ions are generated. These phenomena suppress corrosion and rebuild the passive layer. This means that although the 100 mV shift is lower than the optimum 150 mV, the other factors are also working in our favour. During the initial few months of energizing and commissioning a cathodic protection system it is not necessary to instantly provide total protection as the ionic movements will build up with time. This is discussed further in Sections 6.6.6 and 6.6.7 on energizing and commissioning cathodic protection systems.

The 100 to 150 mV criterion is straightforward to apply and is the most universally agreed criterion. Other control criteria such as the plotting of the applied current against the log of the potential ($E \log I$), absolute potentials, macrocell or null probe current reversals and other potential shifts have been used and are used by some cathodic protection specialists; but there is some controversy about their theory and practice.

An additional criterion concerns the maximum (most negative) permissible potential. If the potential moves more negative than −1100 mV versus a CSE electrode, there is a risk of hydrogen evolution by the reaction discussed at the beginning of the chapter:

$$H_2O + e^- \rightarrow H + OH^- \tag{6.3}$$

The 'nascent' monatomic hydrogen produced is very mobile. It could be trapped at defects and boundaries in the crystal structure of the steel, leading to weakening and embrittlement. This is a negligible risk for conventional reinforcing steel, but a real risk for pre- and post-tensioning steel. There is also a slight risk of bubbles forming at the steel concrete interface and reducing the pull out strength of the reinforcing steel if a lot of hydrogen is evolved.

It is therefore advisable to set an limit of −1100 mV vs. CSE for the instant off potential for all systems. This is more as a convenient upper limit for reinforced concrete structures, but as a rigid limit in the case of prestressing steel if exposed to the cathodic protection current.

6.6 SYSTEM INSTALLATION

Installation of the components of the cathodic protection system is described below. The order given is logical but not necessarily the one used on site where work may proceed in parallel or in a different sequence to fit in with other requirements of site work.

6.6.1 Patching for cathodic protection

Patching for cathodic protection is merely to ensure that there is ionic continuity between the steel and the anode (Figure 6.1). The patch repair material must have the following properties:

1. It must conduct ionically not electronically. There must be no conductive filler such as carbon or a metal. Zinc is used in some patch materials to minimize the incipient anode effect but they must not be used for any electrochemical process repairs. Metal or carbon fibres must not be used either.

2. Conductivity must be low enough to allow current to pass into the steel. A resistivity of 15 kΩ cm at 28 days under saturated conditions has been specified by the UK DoT based on successful use of materials of this resistivity on the Midland Links. There is some disagreement about this specification but it is the only quantitative information available. In North America, some applicators use salt in the concrete to 'even out' the resistivity with the parent concrete. This practice is rarely accepted elsewhere.

Patching for cathodic protection only requires the removal of the corrosion damaged cover, the cleaning of the rebar and its reinstatement. The lack of structural implications and the comparative cheapness of the concrete patching requirements are two important factors in choosing cathodic protection for repair. It must be noted that good quality, low shrinkage materials must be used and well applied. Cathodic protection will not stop a badly applied patch from falling off or developing cracks if badly cured. Figure 5.1 shows the essential elements of a good patch repair for cathodic protection purposes.

Excavated areas may be filled with the same material used to apply the anode, e.g. sprayed concrete on substructures or the concrete overlay on decks. In the case of the titanium mesh, special fixings will hold the anode in position as the concrete fills around it.

6.6.2 Rebar connections and continuity

Electrical connections may be made by self-tapping screws, welding, brazing or thermite welding (see Figure 6.5). The technique used should give a good mechanical and electrical bond without damaging the mechanical properties of the reinforcing steel. Extra bars may need to be welded in if continuity between bars is inadequate. Connections are usually protected with epoxy glue. Cabling may be run through the concrete overlay, over the anode or from the back of the protected surface. Cables are run into junction boxes for splicing and connections. There must be at least two connections per zone for redundancy. There may be separate connections for the reference electrodes (see below).

Rebar continuity is essential to avoid stray currents that can accelerate corrosion. Figure 6.24 shows how an isolated rebar between the anode and the cathode will be cathodic where the current enters the steel and anodic where it exits. This will accelerate corrosion at the anodic site. Although there are few serious cases identified in cathodic protection systems, this is a greater concern for realkalization and desalination systems where the charge density is higher.

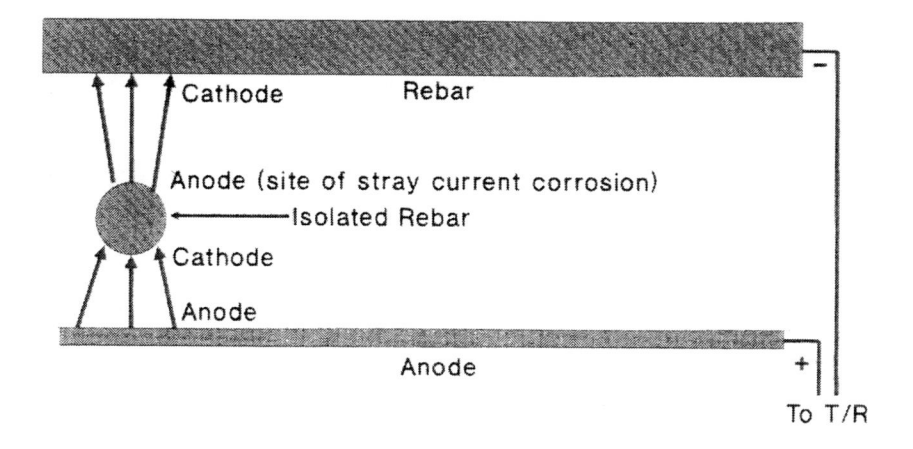

Figure 6.24 Schematic of the effect of a piece of isolated rebar.

6.6.3 Monitoring probe installation

Reference half cell electrodes should be installed in anodic sites without disturbing the concrete around the steel to be measured. There should be at least one electrode per zone and usually more. Other probes such as macrocell or null probes may also be installed but usually in addition to half cells.

6.6.4 Anode installation

Anode installation should be done according to the manufacturer's instructions. These may include details and limits of surface preparation and conditions, such as air and concrete temperatures and moisture content. The anode must not come in direct electrical contact with the reinforcing steel. There must be no metallic fixings into the anode.

The anode system may be a single component, such as flame sprayed zinc, or multiple components such as a titanium mesh with a cementitious overlay or conductive paint coatings with a protective/cosmetic top coat. All anodes require electrical connections to the power supply. The anode and rebar connections should be duplicated for redundancy.

6.6.5 Transformer/rectifier and control system installation

The process of running mains power, installing the T/R and control system and wiring up can be done as a single operation or in stages. All probes and leads should be tested before being connected to the T/R and control system and then tested once the system is installed. T/Rs are factory tested and must comply with national electrical regulations. Lightning arresters, fuses and circuit breakers are usually required for safe and efficient operation. The hardest part of specifying T/Rs is getting the power requirements correct. An underspecified system may not be able to deliver enough power to polarize the steel adequately. An overspecified system may only run at 10% or less of its maximum output. This can make it difficult to control, and inefficient in its operation.

6.6.6 Initial energizing

Anode durability can be strongly affected by the initial energizing and commissioning process. This is particularly true of conductive paint coating anodes where the concrete/anode bond is critical for the long term durability of the system. Passing too much current too quickly through the anode can generate large amounts of acid at the anode as the chloride level near the surface is far higher than at the rebar due to the diffusion process. Acids are formed from oxidizing chlorides to chlorine gas and hydrochloric acid and other reactions. This will attack the concrete paste, consume the anode and reduce the system lifetime, particularly for coating anodes. The problem is less important for mesh and ribbon type anodes. It is therefore essential to energize the system as gently as possible, consistent with proving that the system works and providing protection.

A series of checks should be carried out to ensure that all anodes and probes are correctly wired. Energizing should be at very low current levels initially and should be carried out after the anode system is fully cured and ready according to the manufacturer's instructions.

6.6.7 Commissioning

A series of commissioning tests should be agreed at the design stage. However, the issue of anode durability is still important. A slow increase in current to achieve the minimum required control criteria is recommended. As chlorides move away from the rebar under the influence of the electric field, it is better to under protect in the first months rather than over protect. As long as the system is seen to be capable of delivering adequate current and all test and monitoring areas are polarizing in a sensible manner, with all anodes passing current fairly uniformly, the system should be accepted as functioning correctly.

One option for a commissioning and start up criterion is to achieve 100 mV polarization from the rest potential before energizing to the instant off potential at some agreed commissioning date (typically 28 days after switch on). The achievement of 100 mV depolarization in four hours or some other agreed time period can then be achieved over a more extended period. This should reduce the current requirement during the initial period of high acid generation.

6.6.8 Operation and maintenance

It is essential that impressed current cathodic protection systems are properly maintained. This usually means that a budget, a system of training for personnel and a maintenance schedule are developed by the client, engineer, corrosion specialist and contractor at the initial design and specification stage.

The client may undertake maintenance himself or have an agent do it for him. Even with remote control and monitoring a visual check on the system should be undertaken every month or two and an annual physical inspection conducted leading to repair and maintenance if required.

6.7 CATHODIC PROTECTION OF PRESTRESSED CONCRETE

Most standards state that cathodic protection should not be applied to prestressed concrete structures. This is because of the risk that if the potential exceeds the hydrogen evolution potential, hydrogen embrittlement could occur with potentially catastrophic failure of the steel These problems were discussed in Sections 6.2.1 and 6.5. High strength steel may trap atomic hydrogen which weakens grain boundaries and the crystalline structure. There are several issues that can be considered when considering cathodic protection of prestressed concrete structures:

- Is the steel susceptible to hydrogen embrittlement?
- Will the current reach the prestressing steel (is it in ducts or well behind a reinforcement cage that will absorb the cathodic protection current)?
- Can the prestressing steel be adequately monitored to ensure that the risk of hydrogen evolution is minimized?

In Italy, cathodic protection has been applied to new bridges to protect anchorages, reinforcing steel and exposed tendons in prestressed, post-tensioned bridges (Baldo et al., 1991). The Italian approach is conservative as very modest currents and potentials are needed to protect new structures. However, applying cathodic protection to a new, undamaged structure is thought to admit poor confidence in the initial design by most engineers unless the structure is exposed to a very severe environment such as salt or brine containers. Cathodic protection of prestressed concrete pipelines is also well known, again applied from new, before chlorides have penetrated the concrete.

Initial trials are underway too in the USA to cathodic protection pretensioned structures. Some pretensioned bridge piles have been cathodically protected in Florida using sacrificial anodes (Kessler et al., 1995). This is unlikely to cause overprotection and hydrogen evolution. A Federal Highways Administration contract is underway to carry out trials on prestressed concrete highway structures, using impressed current systems.

The US research will be watched with interest. It has already been noted that a post-tensioned bridge collapsed in Wales a few years ago. The bridge had no waterproofing on the deck, the ducts were poorly grouted and salt got into the joints between segments and corroded the steel. The UK DoT has no idea how many bridges are in a comparable condition in the UK. Further failures are therefore possible. As cathodic protection cannot protect the cables inside ducts on post-tensioned structures, it has limited applicability for protecting such structures.

6.8 CATHODIC PROTECTION OF EPOXY COATED REINFORCING STEEL

Because of the problems of corrosion of fusion bonded epoxy coated reinforcing steel (FBECR) in Florida, (Clear et al., 1995) the Florida DoT and other US researchers have investigated methods of cathodically protecting FBECR. The main problems are establishing the continuity of the steel and the risk of pitting and undercutting corrosion behind the epoxy coating.

A survey of the FBECR structures in Florida that have suffered from corrosion has shown that many have a very high level of continuity. On a survey of 10 bridges, an average of 27% of readings showed electrical continuity (Sagüés, 1994). This may be due to the use of uncoated tie wire in early construction work, and the effect of squeezing bars together at ties.

In order to install cathodic protection, electrical continuity can be established by welding in extra rebars. However, at Florida DoT one approach has been to expose bars in damaged areas, grit blast them clean and apply arc sprayed zinc directly onto the steel and then across the steel surface. This provides galvanizing directly on the steel and SACP to the steel embedded in the concrete. Multiple continuity connections are established by the sprayed zinc.

The problems of pitting and under-coating corrosion are more difficult and are well known in pipeline corrosion where fusion bonded epoxy coatings are frequently applied to the outside of pipelines and then cathodic protection applied to protect the pinholes that inevitably occur. However, these cathodic protection systems are applied from new so no corrosion is established. The FBECR structures are already corroding when cathodic protection is applied. It is therefore possible for corrosion to be established under the coating where the cathodic protection current cannot reach. The small driving voltage of a sacrificial anode system means even less protection or penetration of current than for an impressed current system.

In practice the risks of lack of continuity and of under film corrosion must be accepted. No incidents of ongoing corrosion have been observed on the FBECR structures under cathodic protection after more than five years of trials.

6.9 CATHODIC PROTECTION OF STRUCTURES WITH ASR

Alkali–silica reactivity (ASR) is a condition where certain silicaceous aggregates are susceptible to the alkalinity in the concrete and react to form silica gel. Silica gel is a desiccant used as in chemistry labs and to protect tools and equipment because it absorbs moisture. As it absorbs moisture it swells. The swelling damages concrete when it takes place on aggregate particles in the mix. ASR is characterized by 'map cracking', and weeping of the gel as a white efflorescence from the cracks. One way of controlling ASR at the mix design stage is to limit the maximum alkali content of the mix. This can be done by controlling the chemistry of the cement powder or by blending the cement with pozzolanic additives such as pulverized fuel ash (PFA, also known as fly ash), or ground glass blast furnace slag or microsilica. In these cases marginally susceptible aggregates can be used.

It has been shown above that cathodic protection creates hydroxyl ions and will also attract positive ions such as sodium and potassium to the steel. This will increase the alkalinity around the reinforcing bar. In principle this could cause ASR or accelerate ASR in susceptible mixes. This has been demonstrated in the laboratory. However, there are no recorded cases of ASR being caused or accelerated by cathodic protection in field structures. In a review of field structures by SHRP, a structure with ASR showed no acceleration in the ASR in areas where cathodic protection was applied (Bennett, 1993).

6.10 CHLORIDE REMOVAL

We have already determined that the chloride ion is a catalyst to corrosion (Section 3.2.3). As it is negatively charged, we can use the electrochemical process to repel the chloride ion from the steel surface and move it towards an external anode. This process, called electrochemical chloride extraction, desalination or chloride removal, uses a temporary anode and a higher electrical power density than cathodic protection for a limited time (four to six weeks), but is otherwise similar.

6.10.1 Anode types

The most popular anode is the same coated titanium mesh used for cathodic protection. Instead of embedding it permanently in a cementitious overlay a temporary anode system is used. A proprietary system developed in Norway consists of shredded paper and water sprayed onto the surface to form a wet 'papier mâché'. The mesh anode is then fixed to the surface on wooden batons and a final layer of papier mâché applied. The system is kept wet for the operational period. Figure 6.25 shows an installation underway using a mild steel mesh anode. This is rarely used now as the steel can be selectively consumed during the treatment, leading to uneven current flow and inadequate treatment.

Alternatively a wetting system of a water bath has been applied on bridge decks in the USA. An electrolyte solution circulating system on vertical surfaces using 'blankets' has been used in the SHRP trials on bridges in Florida, Ohio, New York and Ontario (Bennett et al., 1993a) as shown in Figure 6.26. A system of tanks fixed to the concrete surface has been developed in the UK for use with realkalization and desalination as shown in Figure 6.27 (McFarland, 1995).

A 'sacrificial' or consumed anode can be used instead of coated titanium. Copper was tried in the early 1970s trials but copper (or its salts) may accelerate corrosion if it gets into concrete. Steel mesh has been used more recently but has fallen out of favour as it may be completely consumed in some areas before the treatment process is over. It also causes rust staining, although this is not a serious problem as the structure is usually cleaned by water or abrasive blasting after treatment for the application of a protective coating or sealer to prevent further chloride ingress.

6.10.2 Electrolytes

Water is the usual electrolyte, i.e. the medium through which the ionic current flows. This may be dosed with chemicals to keep the solution from going too acidic and etching the concrete, to stop chlorine gas evolution, or to stop ASR (see below). A positively charged corrosion inhibitor has also been added to the solution in some US trials. The inhibitor is described by Asaro *et al.* (1990).

6.10.3 Operating conditions

The very first trials of this system were based on a rapid treatment period of 12–24 hours. Trials were done in Ohio, USA, and lab tests in Kansas Department of Transportation in 1978. However, a more fundamental study carried out by the Strategic Highway Research Program (SHRP) in 1987–92 showed dangers in applying more than 2 A m^{-2} of steel or concrete surface area (Bennett *et al.*, 1993c).

Figure 6.25 Desalination (electrochemical chloride extraction) trial: Burlington Skyway, Ontario, Canada 1989. Acknowledgements Ontario Ministry of Transportation, Eltech, SHRP.

156   *Electrochemical repair techniques*

Figure 6.26 The blanket anode system used in the SHRP desalination/chloride removal procedure.

Ontario Ministry of Transportation carried out a trial of the Norwegian system on Burlington Skyway in 1989. The voltage was kept at about 40 V. The total charge passed was 610 A h m^{-2} of concrete surface area over 55 days giving an average current density of 0.462 A m^{-2}. This structure had a low steel concentration so removal was patchy, very high over the rebars but lower between the bars. They found 78–87% of the chloride removed directly above the rebars and 42–77% of the chloride removed between the rebars (Manning and Ip, 1994).

A trial was carried out in the UK on a section of cross head taken from the corroding substructure of a bridge. The section was removed to a contractor's depot and the sprayed cellulose fibre system applied. The system ran for 92 days passing a total of about 19 565 amp hours

Figure 6.27 The tank anode system used in the Martech realkalization and desalination processes.

charge through approximately 11 m^2 of steel surface. This gave a charge density of 1704 A h m^{-2}, an average current density of 0.77 A m^{-2} and a power density of 25 W m^{-2}.

This work was followed up by laboratory tests on concrete to rebar pull out strength and discussed by Buenfeld and Broomfield (1994).

6.10.4 End point determination

End point determination can be by several means:

1. Point of diminishing returns – resistance goes up, amount of chloride removed goes down, when the current is small and the amount of chloride removed is small, switch off. Switching off for about a week will bring the system resistance down. How much more chloride can be removed by allowing 'rest' periods is not known.
2. Direct measurement – take samples from the concrete and measure the chloride level. When an agreed level is reached, stop. This assumes that good sampling is possible and that samples are representative.
3. Indirect measurement – sample the anode system or electrolyte.

When chloride level is either at a plateau or an agreed level, stop. This also assumes good sampling.
4. Experience of charge density needed – measure charge passed (amp hours per square metre) and when an agreed limit is reached, switch off. The level suggested by SHRP was 600 to 1500 A h m^{-2} (Bennett and Schue, 1993).

In practice a combination of systems is used. Experienced engineers and contractors will know the charge density needed and a trial or test on a core may give a more definite value for a given structure (or element within the structure). Sampling directly and indirectly will show that the system is responding in the expected manner. The point of diminishing returns should be reached soon after the other thresholds.

It is impossible to remove all the chlorides from the concrete by electrical means. The area immediately around the rebar is left almost chloride free but further away there is less effect. This is particularly true behind the steel. Chloride removal will deplete the amount of chloride immediately in contact with the steel and will replenish the passive layer. Somewhere between 50 and 90% of the chlorides are removed. Field data so far (1994) show that this is effective for at least five years but for how much longer is uncertain. The results suggest about 10 years, but only real experience will show.

If large amounts of chloride have penetrated beyond the steel or where cast uniformly into the concrete, then chloride removal may remove much of the chloride in the cover concrete. However, depending on rebar spacing and other factors, very little current will get behind the bars, so much of the chloride beyond the steel will remain. The large reserves of chlorides in the bulk of the concrete may then diffuse back around the steel and the removal process may be very short lived. SHRP experiments on a marine substructure were also discouraging (Bennett et al., 1993a).

However, research shows that the most important aspect of the chloride removal process is the generation of hydroxyl ions, the rebuilding of the protective passive film and the removal of the chlorides immediately around the rebar. SHRP research showed that even with a modest total charge passed and only 50 to 80% chloride removal, and with chloride levels still above the corrosion threshold, treatment will give a very low corrosion rate and very passive half cell potentials which last more than five years without reactivation (Bennett and Schue, 1993). If this is true then we should perhaps call the process 'electrochemical chloride mitigation' and avoid requirements to remove more than 90% of the chloride, as this may not be necessary.

6.10.5 Possible effects

Passing large amounts of electricity through concrete can have effects upon its chemistry and therefore its physical condition. Brown staining around the rebar has been observed on specimens when high current densities (in excess of 2 A m^{-2}) are used (Bennett et al., 1993c). This is an effect on the concrete, not the steel. Current levels are therefore maintained at less than 1 A m^{-2} (usually in the range 0.5–1 A m^{-2}).

There are three known side effects of ECE. The first is the acceleration of alkali silica reactivity (ASR), another is reduction in bond at the steel concrete interface. A third issue is hydrogen evolution. Hydrogen evolution is inevitable with ECE and so this process must not be used on steel used for prestressing.

6.10.6 Alkali-silica reactivity

Research at Aston University in the UK (Page, 1992; Sergi and Page, 1992) and by Eltech Research in the USA under the SHRP program (Bennett et al., 1993c) shows that ASR can be induced by the cathodic reactions (6.2 and 6.3) that generate excess alkali at the steel surface. This is exacerbated by the movement of alkali metal ions (Na$^+$ and K$^+$) to the steel surface under the influence of its negative charge. Some researchers in Japan have suggested that the pH can be so high that the silica gel dissolves, stopping the expansive process.

SHRP undertook a field trial in collaboration with Ontario Ministry of Transportation to see if ASR can be controlled by the application of lithium ions in the electrolyte. These ions (Li$^+$) move towards the rebar under the influence of the electric field. Lithium is known to reduce or stop ASR and has proved effective in lab tests. If a corroding structure is made with aggregates susceptible to ASR, a detailed investigation of its likely reaction to treatment will be required. Monitoring of the Ontario bridge will continue over the next few years to see if the ASR is suppressed (Manning and Ip, 1994).

It has been suggested that if a structure is suffering from ASR and corrosion then corrosion should be the first priority as corrosion will cause the most damage most rapidly. That decision can only be made on a case by case basis.

6.10.7 Bond strength

The effect of current on bond strength of steel in concrete has been a subject for discussion in the technical literature for many years. This is usually with reference to cathodic protection but no effects have been observed in the field on the many hundreds of cathodic protection systems in service for up to 20 years. In most practical applications, the major part of the bond is supplied by the ribbing on the bars so the details of the performance of the steel/concrete interface is irrelevant.

A laboratory study was undertaken to investigate the effect of pull out strength for chloride removal to the Tees Viaduct, a long bridge structure in the north of England (Buenfeld and Broomfield, 1994). Smooth rebars were used in its construction, possibly due to steel shortages at the time of construction.

Some earlier, unpublished tests had shown that ECE can reduce bond strength by as much as 50%. However, careful review of the laboratory data showed that a large amount of charge was passed to obtain this drop in bond, about five times as much as is used normally in an ECE treatment. A very high current density was also used.

The investigation was conducted by Dr Nick Buenfeld of Imperial College Department of Civil Engineering. The author is a consultant to the investigation. Figure 6.28 shows the effect of different levels of charge (current × time) on the pull out strength of smooth bars from cylinder specimens.

These experiments show that bond strength increases as the specimen corrodes (effectively prestressing the concrete by expansive oxide formation). The ECE current appears to eliminate this prestressing effect, although the pull out strength does not fall below the level of the control (Figure 6.28).

If this is in fact the mechanism, then the long-term effect must be

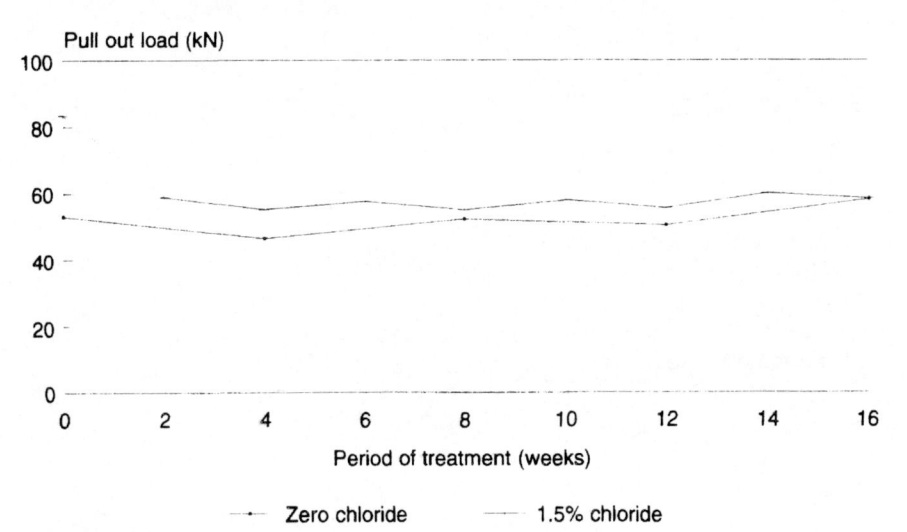

Figure 6.28 Maximum pull out load vs. time at 0.75 A m^{-2} (Buenfeld and Broomfield, 1994).

considered. If corrosion has occurred over a period of five to ten years, then the concrete will creep to accommodate some of the stresses generated by corrosion. If the ECE current then removes the stress it will not be possible for the concrete to revert elastically. Therefore there is a risk of a plane or cylinder of weakness to develop around the bar, leading to a reduction of pull out strength.

A structural analysis of the cross heads of the Tees Viaduct showed them to be adequate but with small safety margins particularly in shear. It was therefore decided to be conservative. Chloride extraction may be considered for the support columns, which are structurally adequate, but the cross heads will be replaced.

6.10.8 Results after treatment: beneficial effects of passing currents through concrete

Comparisons of corrosion rates before and after treatment have consistently shown significant reductions. On the Tees Viaduct field measurements of corrosion rates were in the range 0.33 to 1.66 $\mu A\ cm^{-2}$ with a mean of 0.377 $\mu A\ cm^{-2}$. On a block that was treated, a year later readings ranged from 0.0005 to 0.094 with a mean of 0.0028 $\mu A\ cm^{-2}$. Although no direct 'before and after' measurements were conducted this shows a two order of magnitude difference in corrosion rate between (initially similar) treated and untreated steel in concrete (Broomfield, 1995). Monitoring is continuing.

Resistivity measurement showed large increases after treatment, to over 200 kΩ cm in the field and 5 kΩ cm on untreated lab specimens to 30 kΩ cm after treatment on specimens vacuum saturated with distilled water. The treatment seemed to block pores reducing transport of water, oxygen and chloride ions. This is possibly due to redistribution of calcium hydroxide. There was also an improvement in freeze–thaw resistance (Buenfeld and Broomfield, 1994).

6.11 REALKALIZATION

In equations 3.1 and 3.2 we saw how carbon dioxide reacts with water to form carbonic acid which then reacts with calcium hydroxide to form calcium carbonate. This removes the hydroxyl ions from solutions and the pH drops so that the passive layer is no longer maintained and corrosion can be initiated.

The cathodic reaction (6.2) showed that by applying electrons to the steel we can generate new hydroxyl ions at the steel surface. This regenerates the alkalinity and pushes the pH back up to around 12.

The realkalization process has been patented (Vennesland and Miller, 1992), with a worldwide licensing system for its application. The anodes used are the sprayed cellulose or tank anode systems developed to apply chloride removal. In addition to generating hydroxyl ions, the developers claim that by using a sodium carbonate electrolyte they make the treatment more resistant to further carbonation. The patent claims that sodium carbonate will move into the concrete under electro-osmotic pressure. A certain amount will then react with further incoming carbon dioxide. The equilibrium is at 12.2% of 1 m sodium carbonate under atmospheric conditions:

$$Na_2CO_3 + CO_2 + H_2O \leftrightarrow 2NaHCO_3$$

In laboratory tests they have shown that it is very difficult if not impossible for a treated specimen to carbonate again. Over 80 realkalization treatments (30 000 m^2 of concrete surface area) have been undertaken on structure around Europe over the past few years (1994 figures). The treatment is faster than chloride removal only requiring a few days of treatment.

6.11.1 Anode types

Anode types are the same as for chloride removal. Sprayed cellulose is used in the patented system with a mild steel or coated titanium mesh. The steel is more likely to be used here as the treatment time is shorter and the steel is less likely to be completely consumed.

6.11.2 Electrolytes

Sodium carbonate solution is the preferred electrolyte to give long lasting protection against further CO_2 ingress. However, introducing sodium ions can accelerate ASR so in some cases plain tap water is used. A lithium electrolyte has been proposed but research is still underway on whether it is necessary for realkalization as the concrete starts with a low pH, and whether it is effective.

6.11.3 Operating conditions

In one case a current density of 0.3–0.5 A m^{-2} was applied (at 12 V) to 2000 m^2 of a building in Norway with a treatment time of three to five days. In another case 10–22 V was applied to give a current density of 0.4–1.5 A m^{-2} in 12 days on 300 m^2 of a bridge control tower in Belgium. A further section of 140 m^2 was treated in nine days with a current of 1–2 A m^{-2}. All figures are for concrete surface area. The steel to concrete surface ratio was not given.

6.11.4 End point determination

This is easy for carbonation. A simple measurement of carbonation depth with an acid/alkaline indicator will show when it has been reduced to zero. However, it has been pointed out that the phenolphthalein indicator turns from clear to pink as the pH rises above about 9. This is still an unpassivated condition. Universal indicator or an indicator with a colour change closer to 12 may be required to be sure that alkalinity has been fully restored. The problem with these indicators is that they do not show up well on concrete. Sections 3.1 and 4.8 discuss carbonation and its measurement in detail.

6.11.5 Possible effects

A smaller charge density is applied during realkalization compared to the treatment for chlorides; the risk of damage is therefore lower than for chloride extraction. As mentioned above, ASR may be a risk if sodium carbonate is used as the electrolyte. Sodium carbonate can also cause short-term efflorescence and the high alkalinity after treatment can attack some coatings. Sodium carbonate will attack oil based paints, varnishes and natural wood finishes.

Realkalization has been carried out on prestressed structures, but only on the reinforcing steel. The structure has post-tensioned cables in ducts. These were protected from the current by the ducts and by the fact that they were buried deep in the concrete, well below the reinforcing steel (Miller, 1994). It should be noted that carbonation is a rare problem on prestressed concrete structures as they usually use high strength, high cement content, low water/cement ratio concrete mixes that are highly resistant to carbonation.

6.12 COMPARISON OF TECHNIQUES

It is possible to summarize the advantages and limitations of the techniques so that they can be compared. This is done more thoroughly in the next chapter where Figures 7.1, 7.2 and 7.3 give a breakdown of how to select repairs but the headings below summarize the issues with respect to the electrochemical techniques discussed in this chapter.

6.12.1 Advantages of all electrochemical techniques

1. They treat a large area of a structure, not just patch up immediately corroding areas.
2. They do not give rise to 'incipient anode' problems.

3. Concrete repair to damaged areas is cheaper and easier than for conventional repairs.

6.12.2 Disadvantages of all electrochemical techniques

1. They require specialist knowledge.
2. They cannot be used on prestressed structures, those with ASR, those with epoxy coating, injections or those with poor electrical continuity.

6.12.3 Cathodic protection

1. Advantages: should last 20 years or more with proper maintenance. Proven technology (20 years +), good specifications and standards.
2. Disadvantages: requires permanent power supply. Requires regular maintenance.

6.12.4 Electrochemical chloride extraction

1. Advantages: treatment completed in six to eight weeks with no further maintenance. Temporary power supply can be used, no mains required.
2. Disadvantages: new technology, few standards or specifications available yet. Higher charge density can cause more problems than cathodic protection. Lifetime of treatment not well defined yet.

6.12.5 Realkalization

1. Advantages: treatment completed in two to four weeks with no further maintenance. Temporary power supply can be used, no mains required.
2. Disadvantages: new technology, few standards or specifications available yet

Higher charge density can cause more problems than cathodic protection. Lifetime of treatment not well defined yet.

6.12.6 Costs

It is difficult to give definite cost information as this varies from job to job and country to country. Summaries of costs for cathodic protection are given in Society for Cathodic Protection of Reinforced Concrete (1995) for the UK and in Bennett *et al.* (1993b) for the USA. These techniques are generally only specified if they offer a cost saving to the owner of the structure over its lifetime (Unwin and Hall, 1993).

REFERENCES

AASHTO-AGC-ARTBA (1993) *Guide Specification for Cathodic Protection of Concrete Bridge Decks*, Task Force 29, Federal Highway Administration, Washington, DC.

Asaro, M.F., Gaynor, A.T. and Hettiarachchi, S. (1990) *Electrochemical Chloride Removal and Protection of Concrete Bridge Components (Injection of Synergistic Corrosion Inhibitors)*, Strategic Highway Research Program, SHRP-S/FR-90-002, National Research Council, Washington, DC.

Baldo P., Mason, O., Tettamenti, M. and Reding, J. (1991) *Cathodic Protection of Bridge Viaduct in the Presence of Prestressing Steel: An Italian Case History*, Paper 119, Corrosion 91, NACE International Houston, TX.

Bartholomew, J., Bennett, J., Turk, T., Hartt, W.H., Lankard, D.R., Sagüés, A.A. and Savinell, R. (1993) *Control Criteria and Materials Performance Studies for Cathodic Protection of Reinforced Concrete*, SHRP-S-670 Strategic Highway Research Program, National Research Council, Washington, DC.

Bennett, J.E. (1993) *Cathodic Protection of Reinforced Concrete Bridge Elements: A State-of-the-Art Report*, Strategic Highway Research Program, SHRP-S-337, National Research Council, Washington, DC.

Bennett, J.E. and Schue, T.J. (1993) *Chloride Removal Implementation Guide*, Strategic Highway Research Program, SHRP-S-347, National Research Council, Washington, DC.

Bennett, J.E. and Turk, T. (1994) *Technical Alert: Criteria for Cathodic Protection of Reinforced Concrete Bridge Elements*, Strategic Highway Research Program, SHRP-S-359, National Research Council, Washington, DC.

Bennett, J., Kuan Fuang Fong and Schue, T.J. (1993a) *Electrochemical Chloride Removal and Protection of Concrete Bridge Components: Field Trials*, Strategic Highway Research Program, SHRP-S-669, National Research Council, Washington, DC.

Bennett, J.E., Bartholomew, J.J., Bushman, J.B., Clear, K.C., Kamp, R.N. and Swiat, W.J. (1993b) *Cathodic Protection of Concrete Bridges: A Manual of Practice*, Strategic Highway Research Program, SHRP-S-372, National Research Council, Washington, DC.

Bennett, J., Schue, T.J., Clear, K.C., Lankard, D.L., Hartt, W.H. and Swiat, W.J (1993c) *Electrochemical Chloride Removal and Protection of Concrete Bridge Components: Laboratory Studies*, Strategic Highway Research Program Report, SHRP-S-657, National Research Council, Washington, DC.

Bennett, J.E., Schue, T.J. and McGill, G. (1995) *A Thermal Sprayed Titanium Anode for Cathodic Protection of Reinforced Concrete Structures*, Corrosion 95 Paper No. 504, NACE International, Houston, TX.

Broomfield, J.P. (1995) 'Field measurements of the corrosion rate of steel in concrete using a microprocessor controlled guard ring for signal confinement', in Berke, N.S., Escalante, E., Nmai, C. and Whiting, D. (eds.) *Techniques to assess the corrosion activity of steel reinforced concrete structures*, American Society of Testing and Materials, STP 1276, Philadelphia, PA.

Broomfield, J.P. and Tinnea, J.S. (1992) *Cathodic Protection of Reinforced Concrete Bridge Components*, Strategic Highway Research Program, SHRPC/VWP-92-618, National Research Council, Washington, DC.

Broomfield, J.P., Langford, P.E. and McAnoy, R. (1987) 'Cathodic protection of reinforced concrete: its application to buildings and marine structures', *Corrosion of Metals in Concrete*, Proceedings of Corrosion/87, NACE, Houston, TX, pp. 222–35.

Buenfeld, N.R. and Broomfield, J.P. (1994) 'Effect of chloride removal on rebar bond strength and concrete properties', in Swamy, R.N. (ed.) *Corrosion and Corrosion Protection of Steel in Concrete*, Sheffield Academic Press.

de Peuter, F. (1993) 'New conductive overlay for cathodic protection of reinforced concrete', in Forde, M.C. (ed.) *Structural Faults and Repair*, Engineering Technics Press, Edinburgh. Also Paper 325, NACE Corrosion/93, Houston, TX.

Clear, K.C., Hartt, W.H., McIntyre, J. and Seung Kyoung Lee (1995) *Performance of Epoxy-Coated Reinforcing Steel in Highway Bridges*, NCHRP Report 370, National Cooperative Research Program, Transportation Research Board, National Research Council, Washington, DC.

Geoghegan, M.P., Das, S.C. and Broomfield J.P. (1985) *Conductive Coatings in Relation to Cathodic Protection*, Transport Research Laboratory Report No. P8512100 TRL, Crowthorne, Berkshire, UK.

Hartt, W.H., Chaix, O., Kessler, R.J. and Powers, R. (1994) 'Localised cathodic protection of simulated prestressed concrete pilings in sea water', *Corrosion 94*, Paper 291, NACE International, Houston, TX.

Hime, W.G. (1994) 'Chloride-caused corrosion of steel in concrete: a new historical perspective', *Concrete International*.

Kessler, R.J., Powers, R.G. and Lasa, I.R. (1995) 'Update on sacrificial anode cathodic protection of steel reinforced concrete structures in seawater', *Corrosion 95*, Paper 516, NACE International, Houston, TX.

Manning, D.G. and Ip, A.K.C. (1994) 'Rehabilitating corrosion damaged bridges through the electrochemical migration of chloride ions', in Weyers, R.E. (ed.) *Concrete Bridges in Aggressive Environments*, Philip D. Cady Symposium, pp. 221–44.

McFarland, B. (1995) 'Electrochemical repair of reinforced concrete in the UK', *Construction Repair*, **9** (4), 3–6.

Mears, R.B. and Brown, R.H. (1938) 'A theory of cathodic protection', *Trans. Electrochemical Society*, **74**, 519–31.

Miller, J.B. (1994) 'Structural aspects of high powered electrochemical treatment of reinforced concrete', in Swamy, R.N. (ed.) *Corrosion and Corrosion Protection of Steel in Concrete*, Sheffield Academic Press, pp. 1400–514.

Morgan, T.D.B. (1990) 'Some comments on reinforcement corrosion in stagnant saline environments', in Page, C.L., Treadaway, K.W.J. and Bamforth, P.B. (eds) *Corrosion of Reinforcement in Concrete*, Elsevier Applied Science, London for the Society of Chemical Industry, pp. 29–38.

Pourbaix, M. (1973) *Lectures on Electrochemical Corrosion*, Plenum Press, New York.

Page, C.L. (1992) 'Interfacial effects of electrochemical protection methods applied to steel in chloride containing concrete' in Ho, D.W.S. and Collins, F. (eds) *Rehabilitation of Concrete Structures*, RILEM, Cachan, France.

Sagüés, A.A. (1994) *Corrosion of Epoxy Coated Rebar in Florida Bridges*, Final Report to Florida DOT, WPI No. 0510603, College of Engineering, University of South Florida.

Sergi, G. and Page, C.L. (1992) *The Effects of Cathodic Protection on Alkali–Silica Reaction in Reinforced Concrete*, Contractor Research Report 310, Transport Research Laboratory, Crowthorne, Berkshire, UK.

Stratfull, R.F. (1974) 'Experimental cathodic protection of a bridge deck', *Transportation Research Record*, **500**, Transportation Research Board, Washington, DC.

Society for the Cathodic Protection of Reinforced Concrete (1995) *Status Report: The Cathodic Protection of Reinforced Concrete*, Society for the Cathodic Protection of Reinforced Concrete, Leighton Buzzard, UK.

Unwin J. and Hall, R.J. (1993) 'Development of maintenance strategies for elevated motorway structures', in Forde, M.C. (ed.) *Structural Faults and Repair 93*, University of Edinburgh, Engineering Technics Press, 1, 23–32.

Vennesland, O., and Miller, J.B. (1992) *Electrochemical Realkalization of Concrete*, European Patent Specification No. 0264421.

Whiting D., Nagi, M. and Broomfield J.P. (1995) *Evaluations of Sacrificial Anodes for Cathodic Protection of Reinforced Concrete Bridge Decks*, FHWA-RD-95-041 Federal Highways Administration, Department of Transportation, Washington, DC.

RECOMMENDED READING AND SPECIFICATIONS FOR CATHODIC PROTECTION, DESALINATION AND REALKALIZATION

Cathodic protection

AASHTO-AGC-ARTBA Task Force 29 (1992) *Guide Specification for Cathodic Protection of Concrete Bridge Decks*.

Bennett, J.E. and Bartholomew, J.J. (1993) *Cathodic Protection of Concrete Bridges: A Manual of Practice*, Strategic Highway Research Program, SHRP-S-372, National Research Council, Washington, DC.

Bennett, J. and Turk T. (1994) *Technical Alert: Criteria for the Cathodic Protection of Reinforced Concrete Bridge Elements*, Strategic Highway Research Program, SHRP-S-359, National Research Council, Washington, DC.

Berkeley, K.G.C. and Pathmanaban, S. (1990) *Cathodic Protection of Reinforcement Steel in Concrete*, Butterworths, London.

British Standards Institute (1991) Cathodic Protection Part 1 (1991) *Code of Practice for Land and Marine Applications*, BS 7361.

Concrete Society (1989) *Cathodic Protection of Reinforced Concrete*, Technical Report No. 36.

Concrete Society (1989) *Model Specification for Cathodic Protection of Reinforced Concrete*, Technical Report No. 37.

NACE International RP0290 Standard Recommended Practice (1990) *Cathodic Protection of Reinforcing Steel in Atmospherically Exposed Concrete Structures*.

prEN Draft Standard (forthcoming) *Cathodic Protection of Atmospherically Exposed Reinforced Concrete Structures*.

Society for Cathodic Protection of Reinforced Concrete (1995) *Cathodic Protection of Reinforced Concrete: Status Report*, SCPRC/001.95.

US Department of Transportation/Federal Highway Administration (1988) *Field Inspection Guide for Bridge Deck Cathodic Protection*, US Department of Transportation, Federal Highway Administration, Demonstration Projects Program, Report No. FHWA-DP-34-3, Washington, DC.

Chloride removal

Bennett, J.E. and Schue, T.J. (1993) *Chloride Removal Implementation Guide*, Strategic Highway Research Program, SHRP-S-347, National Research Council, Washington, DC.

NCT (1991) *Specification for Desalination Works*, Norwegian Concrete Technologies, Oslo, Norway.

Realkalization

Meitz, J. and Isecke, B. (1994) *Investigations on Electrochemical Realkalization for Carbonated Concrete*, Paper No. 297, Corrosion 94, NACE International, Houston, TX.
NCT (1991) *Specification for Re-Alkalization Works*, Norwegian Concrete Technologies, Oslo, Norway.
Vennesland, O. and Miller, J.B. (1992) *Electrochemical Realkalisation of Concrete*, European Patent Specification No. 0264421.

7
Rehabilitation methodology

One of the major issues facing any consultant or owner of a structure suffering from chloride or carbonation induced corrosion is what type of repair to undertake. As we have seen from the previous chapters there are coatings and sealants, specialized patch repair materials, options for total or partial replacement, cathodic protection, chloride removal, realkalization and corrosion inhibitors. These can be applied to structures suffering different degrees of corrosion due to chloride attack, carbonation or a combination of the two. Each solution will have implications for the future maintenance requirements and expected service life of the structure.

The practical solution for most owners of individual corroding structures will be to take advice from a civil engineering consultant or a corrosion specialist who works on atmospherically exposed concrete structures. Advice may also be sought from materials suppliers and applicators about their own particular systems and a consensus will be reached about the most effective repair to the structure based on local knowledge, experience and availability of materials and systems.

Many owners of large structures and large portfolios of structures will have developed in-house knowledge. They will have consulted widely and may develop a systematic approach to corrosion rehabilitation.

For highway agencies bridge management systems (BMS) are still in their infancy and are not specifically geared to detailed corrosion assessment or its management, although increasing work is being put into this area. However, as the input into most BMS systems consists mainly of structural and visual survey data, the information on such things as the half cell potentials or chloride contents will not be available from a BMS to make informed corrosion rehabilitation decisions.

The SHRP programme developed a manual and a computer program aimed at providing highway agencies with a way of carrying out lifecycle cost analysis for corrosion rehabilitation of bridges (Purvis *et*

Table 7.1 Selection of compatible treatment strategies

Problem	Overlay	Waterproof	Patching	Cathodic protection	Cl⁻ removal and realkalization	Inhibitors	Reason
Dead load/clearance above deck	Bad	Bad. Needs overlay	OK	Slot anode	OK	OK	Overlays add dead load and cut clearance
Overlay exists on deck	OK replace	Remove and replace (with asphalt)	Difficult	Possible to remove and replace	Difficult	Difficult	Working through overlay gives problems. Removal leaves rough surface
Polymer injection exists	OK	OK	OK	Bad	Bad	Difficult	Treatment cannot reach rebars
No electricity available	OK	OK	OK	Difficult	Possible	OK	Need permanent power supply
Bad rebar continuity	OK	OK	OK	Bad	Bad	OK	Need electrical continuity of rebars
Difficult geometry	Bad	OK	OK	OK	OK	OK	Difficult to lay and finish concrete
ASR problems	Watch alkali content	OK	Watch alkali content	Test for problems	Test for problems	Watch alkali content	Increased alkali could exacerbate ASR
Prestressing in structure	OK	OK	OK	Bad	Bad	OK	Cathodic reactions cause hydrogen embrittlement

al., 1994). The approach makes many assumptions based on the knowledge and experience of the authors and will need further development before it is fully accepted. However, it is probably the first comprehensive systematic attempt to select a rehabilitation method based on engineering judgement rather than subjective judgement.

A detailed recommendation for repair strategies for concrete structures damaged by steel corrosion is given in RILEM (1994).

7.1 TECHNICAL DIFFERENCES BETWEEN REPAIR OPTIONS

Different repair options are suitable for different applications; but these overlap to such an extent that a clear decision on technical or cost grounds is difficult. Equally, no one solution is universally applicable. Table 7.1 gives indications of the compatibilities of repair approaches based on the requirements of the different systems. Giving global information is difficult because there are different approaches in the USA and Europe to concrete repair.

A simplified table of technical issues that may exclude some repair options is given in Table 7.1. This lists some issues such as the design of the structure and other problems that may eliminate certain options or may severely increase the cost of using them.

A repair methodology is given in Figures 7.1, 7.2 and 7.3. They require first, an evaluation of the structure, then a choice between the available options. In all cases there is the 'do nothing' option. This says that corrosion is not too severe and the benefits of waiting (i.e. cost saving and lack of disruption to use of the structure) outweigh the benefits of repairing or rehabilitating now. This option can be attractive for very short times but the structural and safety implications of ongoing corrosion must be fully assessed and understood.

Chapter 8 will discuss how we can evaluate the current condition, develop a condition index and extrapolate forwards to predict future deterioration.

7.2 REPAIR COSTS

Providing general cost guidance is difficult especially as cost information is difficult to find for a generalized approach. Even on a specific project, comparing costs for different repair options can be difficult if not impossible.

This is further complicated in making allowance for inflation effects. The normal US engineering practice is to use a factor of about 3% as the normal difference between inflation and interest rates. However, in the UK the Treasury requires a 7% discount rate as the difference between inflation and the cost of a commercial loan (Unwin and Hall,

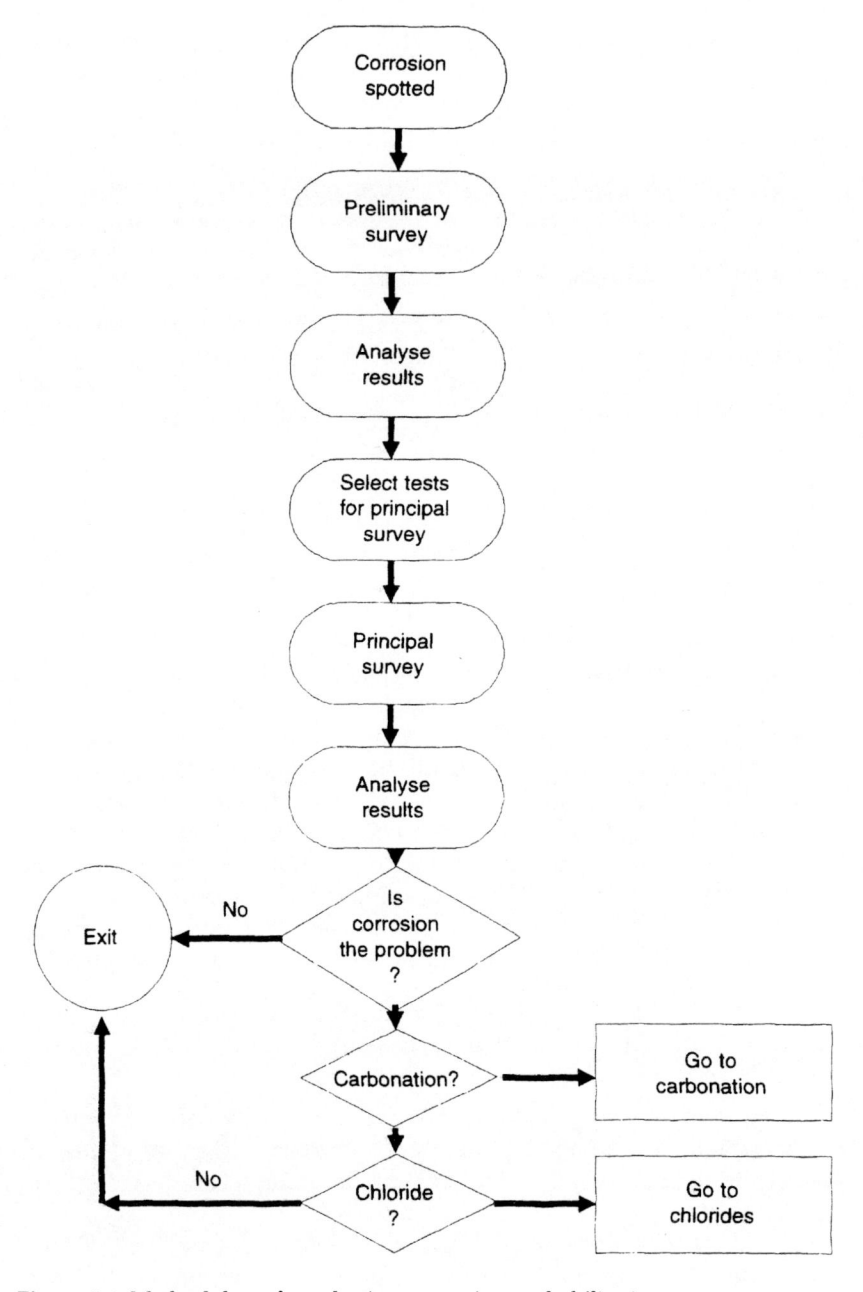

Figure 7.1 Methodology for selecting a repair or rehabilitation.

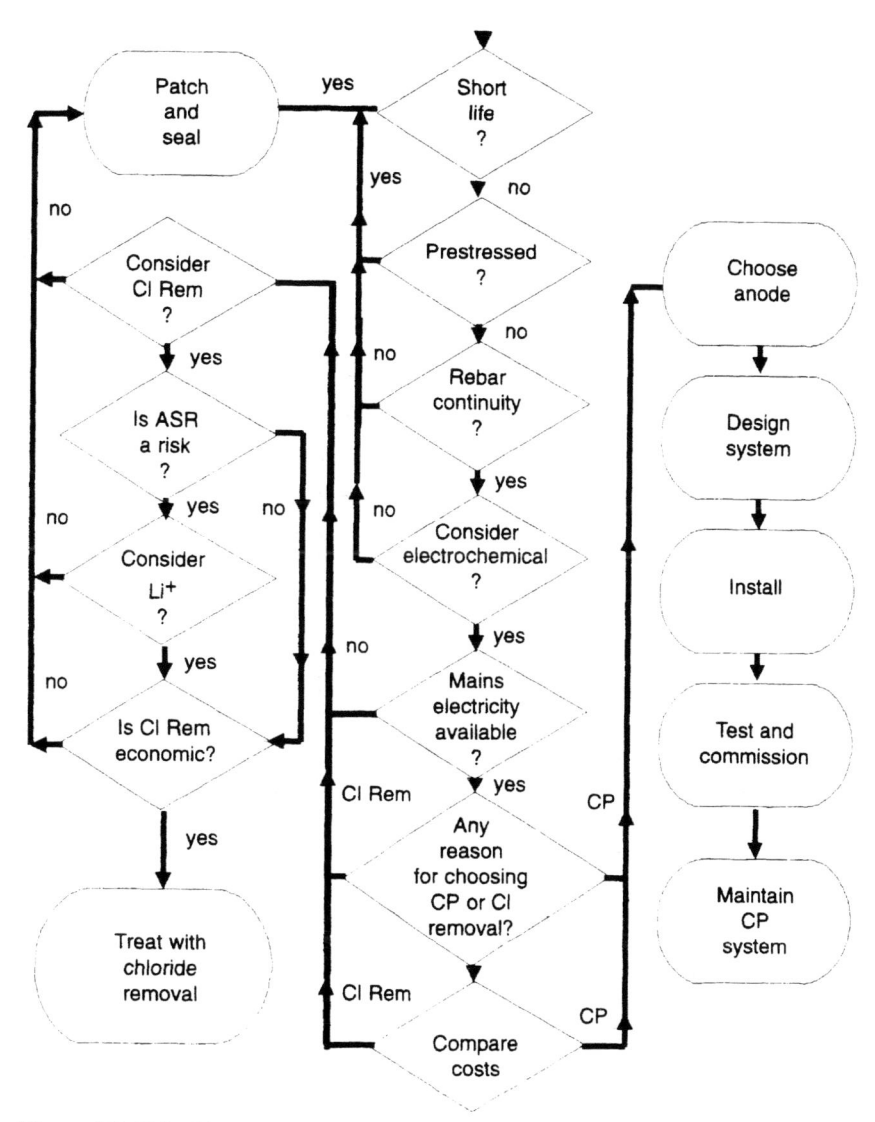

Figure 7.2 Chlorides.

1993). This has a highly distorting effect on all lifecycle costing for UK bridges. It is probably a problem for government buildings and other structures as well.

The most coherent set of comparative data was developed by the SHRP research on repair methods (Gannon et al., 1993). A series of field rehabilitation projects on US bridges was investigated and the data presented in a series of graphs and polynomial equations. These have

Figure 7.3 Carbonation.

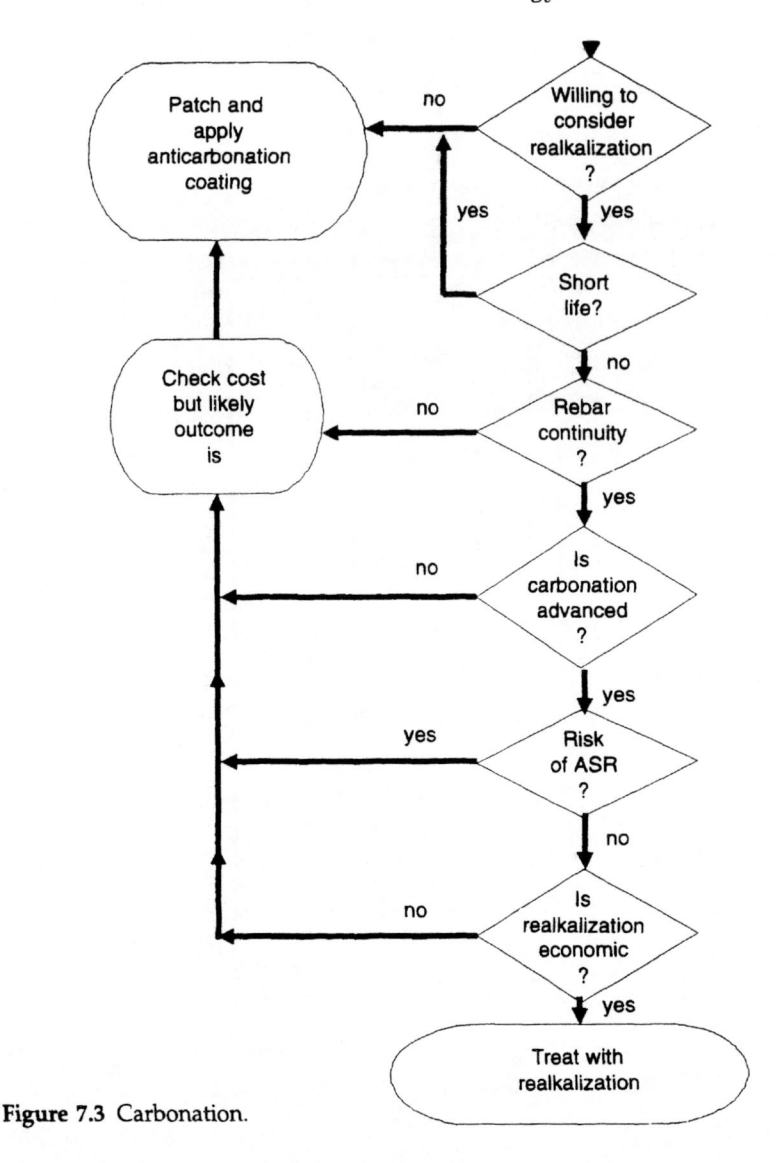

been reduced to some ranges of costs for different areas being treated in Table 7.2. They should be regarded as comparative rather than absolute and giving trends rather than hard numbers.

7.3 CARBONATION OPTIONS

For carbonation the usual options are between patch repairing, usually with an anti-carbonation coating applied afterwards, and realkalization.

Table 7.2 Comparative costings of repair techniques

Method (deck or substructure) (D or S)	0–500 m^2	500–1000 m^2	About 5000 m^2	$\geqslant 10\,000\ m^2$ (L)
Ordinary Portland cement C. part depth (D)	100–880	50–250	50–200	50–200
OPC full depth (D)	50–1300	100–400	100–400	100–400
Latex mod. C overlay part (D)	250–1200	250–350	250–350	250–350
Polymer mod. C overlay part (D)	300–1300	300–500	300–500	300–500
LMC/PMC full (D)	25–175	25–65	30–55	30
Waterproof + asphalt (D)	7–45	6–25	5–12	10
Low slump dense concrete (D)	28–74	28–74	26–68	44–28
Penetrating sealer (D and S)	4–44	4–34	2–36	4–6
Polymethylmethacr. sealer (D and S)	11–52	7–19	7–10	7
Milling (D)	0.5–195	0.5–85	0.5–12	0.5
Asphalt removal (D)	4–37	1–19	1–18	1–4
Thin polymer overlay (D)	45–103	45–103	42–55	42–50
Microsilica overlay (D)	25–108	25–80	25–50	25
Shallow patch OPC (S)	110–1200	110–800	110–400	100
Deep patch OPC (S)	100–5000	100–800	100–700	100
Shotcrete (S)	100–9400	100–3000	100	100
Epoxy coatings (S)	8–80	8–50	8–40	8
Other coatings (S)	8–32	1–25	1–20	1–12
Inhibitors (D and S)	220–2000	120–730	70–500	70–500
Cathodic protection (and Cl^- removal) (D and S)	13–30	9–21	6–14	5–13

Notes: (L) = cost for highest area given (for substructures). Chloride removal has been assumed to be comparable in price to cathodic protection. SHRP US trials gave costs of $130–310 m^{-2}. Inhibitor prices are also from trials only.
Source: Information extracted from figures and tables in SHRP Report SHRP-S-644 *Price and Cost Information* and SHRP Report SHRP-S-360 *Cathodic Protection Manual*. Costs are in 1992 US$ m^{-2}.

Realkalization can only be done with great care and independent checking of the contractor's methods on prestressed concrete structures. There is also a risk of promoting ASR in susceptible structures. These points were discussed in Chapter 6. Realkalization (and all electrochemical techniques) becomes less cost-effective if there are either large numbers of isolated rebars requiring electrical connection together or if there is excessive short circuiting between the reinforcement and the surface that requires checking and removal to prevent shorting to the anode.

A third option is now available in the form of corrosion inhibitors. These can be applied to the concrete surface, to the broken out area to be patch repaired and in the patch repair material.

The rest of this section will discuss the options of patching and coating, realkalization and inhibitor application for carbonation repairs.

7.3.1 Patching and coating

Patching and coating is used for repairing carbonation and chloride induced corrosion damage. They are generally more successful for carbonation. Since the pores in concrete contain significant amounts of water and air, sealing them to stop corrosion is unlikely to be effective once chlorides have penetrated the concrete. This comparatively cheap solution is frequently ineffective once corrosion has started but may give a 'breathing space' for a structure with only a short remaining life or due for major work in a short time. It may also be used to slow or stop the ingress of CO_2 or chlorides before depassivation has occurred.

The effect of moisture reduction was studied by Tuutti (1982) and is discussed in Section 4.11.4. He found a 'critical degree of saturation' for carbonated concrete, where the corrosion rate jumped by almost an order of magnitude. This was at 85% RH. The corrosion rate then rose by two more orders of magnitude to peak at about 95% RH. This implies that if the relative humidity stays below 85% (or 80% for a margin of safety), then the corrosion rate will be low (0.001 to 0.001 $\mu A\ cm^{-2}$). For chloride ingress, the curves are less clear but suggest a significant level of corrosion above 60% RH, peaking nearer 90% RH.

In practice, corrosion monitoring of bridges with sealers and pore blockers or even just sealed joints after years of chloride ingress does show an increase in resistivity and a move toward more passive half cell potentials.

Patching is far more effective for carbonation than for chlorides. This is because restoring the alkalinity is all that is necessary to stop corrosion in carbonated concrete and an anti-carbonation coating will then slow carbon dioxide ingress. For chlorides, patch repairs are only effective if chloride ingress is local and the chlorides can all be removed. Just patching up damaged areas is a short-term palliative not a long-term rehabilitation. The discussion in Section 6.1 on patching shows how the 'incipient anode effect' can actually accelerate corrosion around a patch. This phenomenon is usually observed in chloride conditions rather than carbonation, which is why patching is more effective in carbonated rather than chloride contaminated structures. Contaminated or carbonated concrete must be removed from all around the steel. In some cases temporary structural support may be needed during the repair process.

For accurate quotations and estimates it is important to define the areas and volume of repair accurately. One of the biggest problems encountered in carrying out patch repairs is that the amount of patching is underestimated. This is usually for two reasons, the inaccuracy of the 'sounding' technique for delimiting areas of delamination and the growth of those areas in the time between surveying and the repair contractor carrying out the work. The next chapter discusses how the estimate can be revised with time, especially if there are excessive delays.

7.3.2 Why choose realkalization?

As stated above, patching and applying an anti-carbonation coating is much more effective than patching and coating for chloride attack. It was the only repair option for carbonated structures until the early 1990s. The problems of continued outbreaks of corrosion that had led to the development of cathodic protection for chloride contaminated structures had not manifested themselves for carbonated structures. However, in the late 1980s realkalization was developed and it is an increasingly popular method of rehabilitating structures where carbonation is a significant and recognized problem.

The extent to which carbonation has reached the rebar and the requirements for patch repairing to restore alkalinity will determine whether realkalization is preferred either because it is more economic or in other cases to avoid the noise, dust and vibration required for extensive patch repairing. As the patches can be quite shallow (see Figure 5.1), temporary structural support may be avoided as the concrete behind the steel is not removed. The technique is becoming popular in Europe and the Middle East. In North America very little attention is paid to carbonation. Whether this is a difference in climate, concrete or perception is not clear. The amount of attention dedicated to the chloride problem on highway bridges may have led to the minimizing of the perception of the problem of carbonation in North America.

Figure 7.4 shows the growth in realkalization treatment from the development of the patented process in 1987. The figures for chloride removal have been included but are a very small proportion of the totals in the period shown.

If realkalization is as effective as claimed then the choice between realkalization and patching and coating is a question of convenience

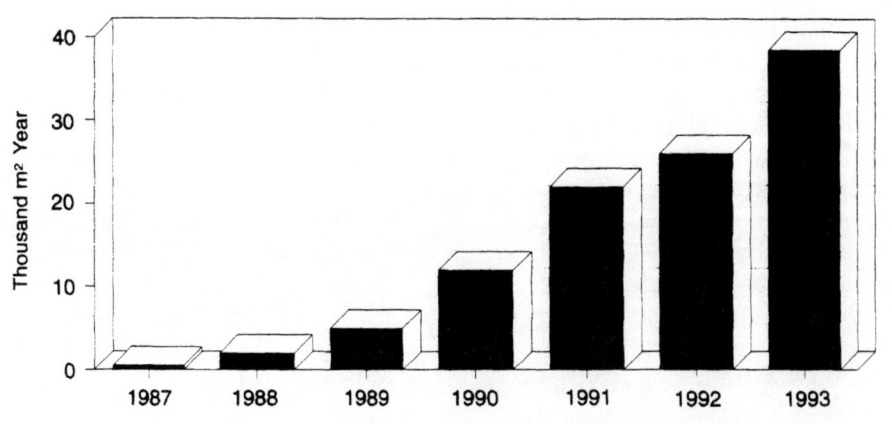

Figure 7.4 NCT figures for growth of treatments by realkalization and desalination.

and cost, together with a realistic appraisal of the effectiveness of anti-carbonation coatings.

The patentees and the licensees of the realkalization system claim that:

- It is financially competitive with the alternatives.
- Vibration and noise are greatly reduced during the (shortened) concrete repair phase.
- All of the surface is treated.
- Guarantees are offered.
- In many cases a fixed price can be offered with no increase of the final bill for extra repair work.

The main questions to be asked when considering this option are:

- How long will it last? Is that time compatible with the service life requirements of the structure?
- Is repeat treatment feasible and acceptable if carbonation resumes at the end of the service life of the treatment (say after 10 years)?
- Are there any problems with my structure that could interact with the realkalization process?

The main requirements for realkalization are:

- electrical continuity of the steel;
- reasonable level and uniformity of conductivity of the concrete when wet;

- no metallic (electronic) short circuits to the surface are present;
- no risk of causing or accelerating alkali–silica reactivity;
- no risk of hydrogen embrittlement of prestressing steel;
- availability of electrical power for the treatment period;
- no electrically insulating layers between the surface and the reinforcing steel.

These requirements can be checked as an extension of the condition survey or as a specific feasibility survey/study. If there are problems with continuity or shorts or high resistance patch repairs then these can often be dealt with but the cost should be analysed to ensure that the process is still cost effective.

The two major problems are ASR and prestressing. Careful studies should be carried out if ASR is a risk. This was discussed in the previous chapter as a problem for cathodic protection (Section 6.9), chloride removal (Section 6.10.6) and realkalization (6.11.5). Although the electrical current density and charge levels are lower than for chloride removal, there is still a risk that the hydroxyl ions generated at the steel will increase the alkalinity around the aggregate particles and cause ASR. This can be tested in the laboratory by taking cores and subjecting them to the current levels to be used in the realkalization process.

The problem of hydrogen embrittlement and prestressing steel has been discussed in Section 6.7 for cathodic protection, 6.10.5 for chloride removal and 6.11.5 for realkalization. The realkalization process applies 20 to 50 V DC between the anode and the steel. It must therefore send the steel potential well beyond the level needed for hydrogen evolution.

Great care must therefore be exercised when applying realkalization to structures containing prestressing steel under load. Very careful monitoring must be carried out to ensure that the steel is completely shielded from the risk of hydrogen evolution or the steel must be carefully tested to ensure that it has no susceptibility to embrittlement from nascent hydrogen. These issues are discussed in Miller (1994).

Due to the problems with ASR and prestressing steel, realkalization would probably not be used as the alternative of patch repairing and coating would provide more cost effective solutions. In any case, carbonation is most unlikely to take place in prestressed structures due to the high cement contents and high cover in such elements. If carbonation had reached prestressing steel in such structures there would probably be concerns about the strength of the concrete.

7.3.3 Why choose corrosion inhibitors?

Although nitrites have been on the market as corrosion inhibitors for a number of years there are now formulations of these and other inhibitors being tested for treating repairs. These include monofluorophosphate (MFP), vapour phase inhibitors and other proprietary formulations.

When considering the use of inhibitors the following questions should be asked:

- Is there a way of ensuring that the inhibitor will reach the rebar in a reasonable time?
- What level of the inhibitor is needed and can be achieved?
- How long will the treatment last?
- Are there any potentially deleterious effects?

The question of inhibitor penetration is one that is not yet fully resolved for any of the inhibitors. Methods of getting inhibitors into the concrete were discussed in Section 5.6. Obviously if the inhibitor is sprayed directly on to the rebar in a repaired area, or if applied into drilled holes or slots cut into the surface, or to a milled surface before installing an overlay then there will be more rapid access to the steel than for a simple application to the concrete surface.

If the concrete is dry at the time of application and has an open pore structure then inhibitor ingress will be more rapid than for a dense concrete which is already full of moisture. Different inhibitors will be transported through the concrete in different ways depending on the size of the molecules and the phase (liquid or vapour).

Most inhibitors have been tested for any deleterious effects but checks should be made for the following:

- odours, particularly with vapour phase inhibitors used indoors;
- staining, efflorescence or problems with coating the concrete after treatment;
- effect of inhibitor admixtures on patch repair mixes or overlays, e.g. flash setting with calcium nitrite, retardant effects of other inhibitors.

At the time of writing (1996) any use of inhibitors to rehabilitate corrosion damaged concrete should be considered experimental. Trials on cores and test areas are recommended with follow up studies using corrosion rate measurement before and after application, with control areas for comparison.

7.4 SUMMARY OF OPTIONS FOR CARBONATION REPAIRS

Table 7.3 summarizes the merits and limitations of the three rehabilitation procedures discussed. As said earlier, there is no single ideal solution to the problem. It is a matter of cost, condition of the structure and the merits and limitations of the techniques as to which to include

Table 7.3 Comparison of techniques for carbonation repair

Technique	Effectiveness	Limitations	Side effects/problems
Patch repair and anticarbonation coating	Yes where patched	Costs can rise if extra repairs needed	Structural support may be needed
Realkalization	Yes across treated area	May need repeat after about 10 years	Can aggravate ASR. Embrittlement of prestressing steel
Inhibitors	Yes across treated area	Rate of penetration to steel not known	Low dosing could cause pitting. Still experimental

or eliminate from the list of options, including the ubiquitous option of 'do nothing', which must always be considered.

7.5 CHLORIDE OPTIONS

If chlorides are the cause of corrosion then patching (including overlaying and encasement), cathodic protection and chloride removal are the three most likely options, with inhibitors a possibility. As seen in the previous section, all techniques have different merits and limitations. This section will attempt to give comparisons.

7.5.1 Patching and sealing

Some degree of patching will probably be required whatever other repair is required. The merits and limitations of patch repairs have been discussed in the previous chapter along with the needs for different depths of repair for cathodic protection or chloride removal. It may also be possible to use less stringent repairs if corrosion inhibitors are being applied in the patch repair material and to the concrete. The issue of corrosion induced around the patch has also been discussed (the incipient anode, Figure 5.5).

The limitation of trying to seal concrete after chloride has started to corrode the rebar has also been discussed. Coatings and sealers may slow ongoing chloride ingress but as chlorides are catalysts to corrosion sealing is unlikely to stop corrosion unless it can be a guaranteed total barrier to moisture or oxygen ingress. Again, however, it may slow corrosion in structures that are only just activated, as discussed in Section 7.3.1.

The amount of patch repairing necessary and its underestimation are discussed in Section 7.4 above under carbonation repairs.

There are two exceptions to the universal need to carry out some patch repairs as part of the rehabilitation process:

1. Chlorides have been detected and a preventive treatment such as cathodic protection is to be applied before delamination starts.
2. The level of cracking and spalling is acceptable and cathodic protection or chloride removal is to be applied as a 'holding' treatment to prevent further deterioration

7.5.2 Why choose cathodic protection?

This technique has now been used for more than 20 years on reinforced concrete structures. However, anode systems are still developing and most cathodic protection systems on reinforced concrete structures are less than 10 years old. No deleterious effects have been found over the 20-year period apart for one anode system that failed after about five years. There have, inevitably been problems and failures of individual installations with the many hundred that have been installed across the world. Cathodic protection has been applied to concrete bridges in the USA since 1973 and in the UK to buildings and other structures since 1988. It has been stated by the US Department of Transportation that cathodic protection is the only method of stopping corrosion across the whole structure, regardless of chloride content. This statement was made before chloride removal was available, and the advocates of chloride removal would no doubt make the same claim for that technique too.

There are now many national, international and proprietary specifications on cathodic protection for steel in concrete (see the bibliography at the end of Chapter 6). Like chloride removal, cathodic protection can only be applied with great care if there is prestressing in the structure. There is a considerable amount of ongoing research in this area. Careful investigation is recommended for structures suffering from ASR. There are also extra installation costs if there are either large numbers of unconnected rebars or if there is excessive short circuiting between the reinforcement and the surface.

A cathodic protection system requires steel continuity, no shorts to the surface, a power supply for the life of the system and continuous monitoring and maintenance.

The main advantages of cathodic protection for chloride attack are as follows:

- It prevents corrosion damage across the whole of the area where the cathodic protection system is applied.

Chloride options

- Corrosion should not recur throughout the life of the system.
- It has a long (20-year) track record of successful application to bridges and other reinforced concrete structures above ground, with hundreds of structures and hundreds of thousands of square metres treated.
- There are established, non-proprietary specifications for its application.
- There are anode systems suitable for most applications.
- Competitive tendering is routine in most countries.
- Lifecycle cost analysis has shown it to be cost-effective for a wide range of structures and conditions.

The following checklist should be considered for cathodic protection and is similar to that given above for realkalization:

- Is there electrical continuity of the steel?
- Is there a reasonable level and uniformity of conductivity of the concrete?
- Check for the presence of metallic (direct electric) short circuits to the surface.
- Check that there is no risk of causing or accelerating alkali silica reactivity.
- Check that there is no risk of hydrogen embrittlement of prestressing steel.
- Check the availability of electrical power for the life of the structure.
- Ensure that there is a management system in place for monitoring and controlling the system.

Continuity and short circuits can be remedied prior to installation if necessary. At present all applications of cathodic protection to structures with prestressing are experimental with the exception of the 'cathodic prevention' applied to new bridges in Italy to keep corrosion from initiating. This uses very low level currents and voltages with special control systems to prevent hydrogen embrittlement.

7.5.3 Why choose chloride extraction?

Cathodic protection has been described as the only solution to chloride induced corrosion that can stop (or effectively stop) the corrosion process. The problems with cathodic protection are its requirements for a permanent power supply, regular monitoring and maintenance.

Similarly chloride extraction can stop corrosion across the whole structure and has the advantage that, like a patch repair, it is a one off treatment. A generator can be brought in for the duration of the treatment so mains power is not needed. There is no long-term maintenance need but the system does treat the whole structure.

The disadvantage is its unknown duration of effectiveness. We cannot remove all the chlorides from the concrete. If we can stop further chloride ingress then the treatment may be effective for many years (10 to 20). If chlorides are still impinging on the structure then the time to retreatment will be shorter (or may require a coating or sealant).

Chloride removal cannot be applied to prestressed structures due to the risk of hydrogen embrittlement. Work is progressing on its application to structures suffering from ASR but this is in the early development stage at the moment. As stated earlier there must be electrical continuity within the reinforcement network for any of the electrochemical techniques to be applied.

The main advantages of chloride removal are as follows:

- It prevents corrosion damage across the whole of the area treated.
- There are a number of different anode systems suitable for different applications.
- Lifecycle cost analysis has shown it to be cost effective for a wide range of structures and conditions.
- It does not require a permanent power supply. A generator or similar temporary source can be used.
- It does not require regular maintenance or monitoring after the initial treatment (although checks on reinitiation of corrosion may be needed on a yearly or two-yearly basis).

The following checklist should be considered for chloride removal and is similar to that given above for realkalization and cathodic protection:

- Is there electrical continuity of the steel?
- Is there a reasonable level and uniformity of conductivity of the concrete?
- Are there are metallic (direct electrical) short circuits to the surface?
- Is there a risk of causing or accelerating alkali silica reactivity?
- Is there a risk of hydrogen embrittlement of prestressing steel?

7.5.4 Other chloride repair options

The other chloride repair options include:

- inhibitors (see Section 5.6);
- concrete overlays (used on North American highway bridges according to specifications developed by highway agencies);
- sealers (more appropriate before corrosion is initiated);

Table 7.4 Comparison of techniques for chloride repair

Technique	Effectiveness	Limitations	Side effects/problems
Patch repair and coating	Yes where patched	Costs can rise if extra repairs needed	Structural support may be needed
Patch repair and overlay/encase	Depends on chloride level and amount of concrete removed	Cost rise if more delaminations found	Can go right through decks
Cathodic protection	Yes across treated area	Needs monitoring and maintenance	Can aggravate ASR, embrittlement of prestressing steel
Chloride removal	Yes across treated area	Lifetime not known	As for CP above
Inhibitors	Yes across treated area	Rate of penetration to steel not known	Low dosing could cause pitting. Experimental still

- waterproofing membranes (used before corrosion has initiated);
- coatings, barriers and deflection systems.

The last three systems are best applied well before corrosion has initiated. As the chloride profile within the concrete approximates to a square root relationship with depth of steel from the surface, there will be considerable reserves of chloride in the cover concrete to push the concentration at rebar level above the corrosion threshold even if the supply is completely cut off. It may be advisable to calculate the redistribution of chlorides to see if and when redistribution will lead to depassivation.

Table 7.4 gives a brief summary comparison of rehabilitations for chloride induced corrosion of reinforcing steel.

7.6 SUMMARY

We have seen that the selection of a suitable rehabilitation technique can be based on technical considerations and cost (preferably whole life costing rather than just initial cost). The technically unacceptable can be excluded and a short list of suitable rehabilitations can be drawn up. Lifecycle cost analysis techniques have been described in the literature to calculate the optimum time and the optimum repair on bridges but are based on a number of assumptions and estimates, including cost estimates for different repair strategies. A direct comparison of quotations for a given structure is probably the best present state of the art although it is important to accurately define the corrosion conditions so that accurate bills of quantities can be drawn up.

REFERENCES

Gannon, E.J., Cady, P.D. and Weyers, R.E. (1993) *Concrete Bridge Protection and Rehabilitation: Chemical and Physical Techniques: Price and Cost Information*, Strategic Highway Research Program, SHRP-S-664, National Research Council, Washington, DC.

Miller, J.B. (1994) 'Structural aspects of high powered electrochemical treatment of reinforced concrete', in Swamy, R.N. (ed.) *Corrosion and Corrosion Protection of Steel in Concrete*, Vol. 2, Sheffield Academic Press, pp. 1499–511.

Purvis, R.L., Babaei, K., Clear, K.C. and Markow, M.J. (1994) *Life-Cycle Cost Analysis for Protection and Rehabilitation of Concrete Bridges Relative to Reinforcement Corrosion*, Strategic Highway Research Program, SHRP-S-377, National Research Council, Washington, DC.

RILEM Draft Recommendation 124-SRC (1994) 'Repair strategies for concrete structures damaged by steel corrosion', *Materials and Structures*, **27** (171), 415–36.

Tuutti, K. (1982) *Corrosion of Steel in Concrete*, Swedish Cement and Concrete Research Institute, pp. 84–9.

Unwin, J. and Hall, R.J. (1993) 'Development of maintenance strategies for elevated motorway structures', in Forde, M.C. (ed.) *Structural Faults and Repair 93*, Vol. 1, Engineering Technics Press, Edinburgh, pp. 23–33.

8

Understanding and calculating the corrosion of steel in concrete

Corrosion of steel in concrete can be modelled as a three-stage process. The first, usually called the initiation stage, is the diffusion of CO_2 or chlorides to the steel to cause depassivation. The second stage is the 'activation stage'. This is a less clear cut phenomenon as more of the rebar network starts to corrode, and the rust products are formed. The third stage is deterioration as cracking and spalling occur. Eventually a situation is reached which is defined as the end of the functional life and rehabilitation must take place by that point. The process is illustrated in Figure 8.1.

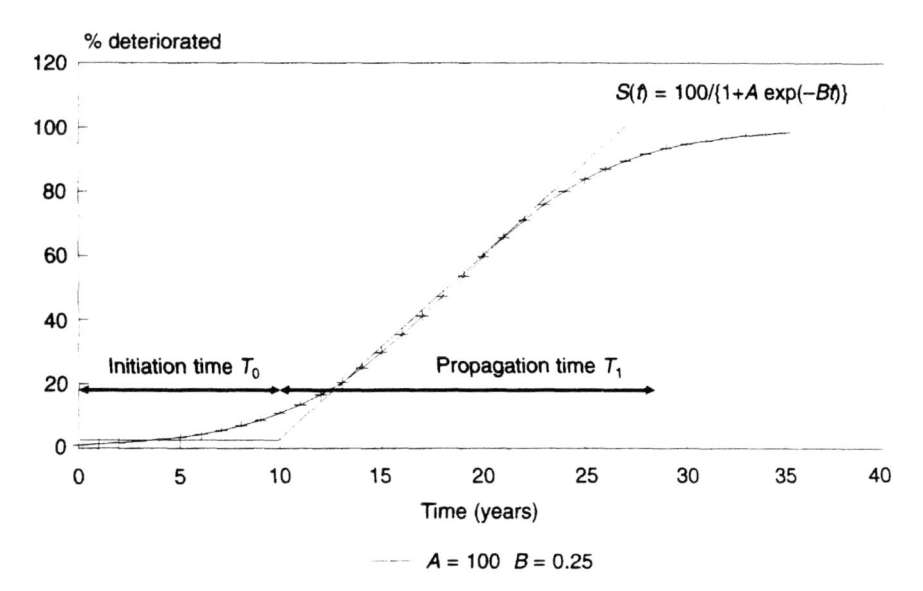

Figure 8.1 Deterioration curve and condition indices.

Curve: $S(t) = 100/\{1+A \exp(-Bt)\}$, with $A = 100$, $B = 0.25$. Initiation time T_0, Propagation time T_1.

8.1 INITIATION TIME T_0, CARBONATION INDUCED CORROSION

Carbonation and chloride induced corrosion are both generally considered to be diffusion based phenomena. The rate of progress of carbonation is given by the equation:

$$d = At^n \qquad (8.1)$$

where d is the carbonation depth in millimetres, A is the A coefficient, t is time (years) and n is an exponent, usually $= \frac{1}{2}$. This is a simple derivation of the diffusion law that says the rate of diffusion is inversely proportional to the thickness being diffused through:

$$dx/dt = 1/kx$$

where x is the thickness and t is time. Therefore $dt = kx\,dx$. By integration $t = Kx^2 + K_0$.

In Scandinavia, the constant of integration K_0 is used to reflect any initial maltreatment that can lead to initial carbonation. For instance an average quality concrete that was deshuttered too early and may have had slight surface freezing can give a $K_0 = 10$ mm. A properly treated concrete has $K_0 \leqslant 1$ mm (Miller, 1995).

Assuming that there is no carbonation at time zero, $x = K'T$. This is based on the assumptions of steady state, i.e. constant CO_2 content at the surface and constant, uniform conditions within the concrete.

This is a very simple equation to use if we can derive a value for the constant. However, we do not always observe this straightforward behaviour. There are several reasons for deviation from the parabolic relationship. These include changing conditions such as humidity and temperature that will change the coefficient, giving apparent deviation from the t behaviour. Also the concrete quality changes with depth and across the structure so the diffusion constant will vary.

This leads to the generalized equation:

$$x = At^n \text{ where } 0 \leqslant n \leqslant 1$$

A number of empirical calculations have been used to derive values of A and n based on such variables as exposure conditions (indoors and outdoors, sheltered, unsheltered), 28-day strength and water/cement ratios as shown in Table 3.1. The easiest solution for a given structure is to take some measurements of carbonation depth, assume $n = \frac{1}{2}$ and calculate A. This can be used to predict the rate of progression of the carbonation front. The time taken to reach the steel can then be estimated and the rate of depassivation calculated.

8.1.1 Parrott's determination of carbonation rates from permeability

For a new concrete mix or structure, the prediction of carbonation rate is complicated by the lack of data to extrapolate. In a series of papers (Parrott and Hong, 1991; Parrott, 1994a and b), a methodology was outlined for calculating the carbonation rate from air permeability measurements with a specific apparatus. Parrott analysed the literature (Parrott, 1987), see Section 3.1.1, and suggested that the carbonation depth D at time t is given by:

$$D = ak^{0.4}t^n - c^{0.5}$$

where k is air permeability (in units of 10^{-16} m^2), c is the calcium oxide content in the hydrated cement matrix for the cover concrete, $a = 64$. k can be calculated from the value at 60% relative humidity by the equation:

$$k = mk_{60}$$

where m is $1.6 - 0.00115r - 0.0001475r^2$ or m is 1.0 if $r < 60$, n is 0.5 for indoor exposure but decreases under wetter conditions to:

$$n = 0.02536 + 0.01785r - 0.0001623r^2$$

To measure the concrete cover depth required to prevent carbonation from reaching the steel it is therefore necessary to measure the air permeability and the relative humidity, and then calculate D. This can be done with a proprietary apparatus developed by Parrott and available commercially.

8.2 CHLORIDE INGRESS RATES (INITIATION)

Chloride ingress is more complex as we are dealing with atmospherically exposed structures exposed to variable chloride concentrations. Also, there is no 'front' that moves through the concrete but a chloride profile or gradual increase in concentration that builds up in the concrete with depth and with time.

The usual form of the diffusion equation used is Fick's second law:

$$d[Cl^-]/dt = D_c \, d^2[Cl^-]/dx^2$$

where $[Cl^-]$ is the chloride concentration at depth x at time t and D_c is the diffusion coefficient (usually of the order 10^{-8} cm^2 s^{-1}).

The exact solution to this equation is:

$$(C_{max} - C_d)/(C_{max} - C_{min}) = \mathrm{erf}[(x)/(\{4D_c t\}^{1/2})] = u \qquad (8.2)$$

where u is the concentration ratio for each depth d below the maximum concentration C_{max} at or near the surface, C_{min} is the background level of cast in chloride (assumed to be evenly distributed) and erf( ) is the error function described by the argument in brackets.

In atmospherically exposed concrete there is no easy initial number C_{max} to use for surface concentration as the chlorides at the surface can vary with time and conditions from zero to 100% depending upon wetting, drying, evaporation, wash off, etc. It is therefore common practice to discard the first 5 mm or so of a chloride profile sample and take the next increment, around the 10 mm depth, as a constant initial, pseudo-surface concentration. If this is done then diffusion calculations must use the depth from the sampling depth, not from the surface.

8.2.1 The parabolic approximation

A simplified method of calculating the initiation time for chloride attack is to look at the progress of the 'chloride threshold' through the concrete. By taking samples with depth it is possible to fit a parabolic curve to the chloride concentration (or more simply to fit a straight line to a plot of depth versus the square root of chloride concentration) and to find the depth at which the concentration is 0.4% chloride by weight of cement (or 1 lb per cubic yard or whatever threshold is chosen). Its rate of progress through the concrete can then be predicted using the simple diffusion equation 8.1. Estimates of the time to reach the rebar can then be made (or the more complex equation 8.2 can be used). However, any background concentration of chloride must be allowed for.

An additional complication comes from the recent observation that the diffusion constant appears to change with time. It is therefore important to only take measurements on 'mature' concrete as the diffusion rate may be higher in the first few years.

In a report to the Nordic Concrete Research Group, Poulsen (1990) pointed out that within defined mathematical limits the error function expression above for the diffusion coefficient can be approximated to the simple parabolic function. He used the following expression for Fick's second law:

$$C_{(x,\ t)} = C_i + (C_s - C_i)\ \mathrm{erfc}\ [x/(4tD_0)^{1/2}]$$

where $C_{(x,\ t)}$ is the chloride concentration at time t and depth x (the profile), C_i is the initial chloride concentration, C_s is the surface chloride concentration, erfc(z) is the error function complement (see below) and D_0 is the diffusion constant.

By using the approximation $\text{erf}(z) = (1 - z/\sqrt{3})^2$:

$$C_{(x,\ t)} = C_i + (C_s - C_i)[1 - x/(12tD_0)^{1/2}]^2 \tag{8.3}$$

for $0 \leqslant x \leqslant (12tD_0)^{1/2}$ and for $x \geqslant (12tD_0)^{1/2}$ then $C_{(x,\ t)} = 0$. The equation 8.3 is in the form $y = mx + c$.

If a series of incremental drillings is taken, then a plot of square roots of the concentration change $(C_s - C_i)$ versus distance into the concrete can be made. A straight line can be fitted, ignoring any surface effects due to washing out of chlorides or concentration due to evaporation of water leaving excessively high levels of chloride behind. The error limits should be checked to exclude extreme values.

The straight line graph can be used to calculate the diffusion coefficient and the rate of movement of any required chloride threshold in mm $\sqrt{\text{yr}^{-1}}$. The process is recommended for use by Germann Instruments with their RCT profile grinder and is fully described with examples in their instruction and maintenance manual (German Instruments, 1994).

8.2.2 Sampling variability for chlorides

Inevitably an inhomogeneous material like concrete will show variation in the chloride content unless very large samples are taken due to the variation in the ratio of paste to aggregate. There will also be some variation due to the local ability of the matrix to resist chlorides. Further, in many atmospherically exposed structures there will be very large differences in chloride content due to the differences in exposure. For instance, on a bridge substructure there will be areas of water rundown that are exposed to very high levels of chlorides while adjacent areas are comparatively unaffected. Chloride laden water may pond at some sites (e.g. on the top of beams, especially if they are horizontal). At the bottom of the beam the water may evaporate leaving the chloride behind.

This can make it very difficult to sample consistently or identify typical environments. It is made even more difficult if intervention has started, e.g. by applying gutters under the joints, building up concave cross heads, etc. where water had previously ponded. These early interventions can prevent easy identification of areas susceptible to high or low chloride ingress.

8.2.3 Mechanisms other than diffusion

It is important to recognise that diffusion is not the only transport mechanism for chlorides in concrete, particularly in the first few millimetres of cover. There may be several mechanisms moving the chlorides including capillary action and absorption as well as diffusion. Rapid initial absorption occurs when chloride laden water hits very dry concrete. In many circumstances these will only affect the first few millimetres of concrete. If so then the expedient of ignoring the first few millimetres of drillings and then calculating diffusion profiles will work. If the cover is low, the concrete cycles between very dry and wet or the concrete quality is low then the alternative transport mechanisms may overwhelm diffusion, at least to rebar depth.

This rapid initial absorption of chlorides may help to explain the over estimate of the diffusion rate often made when predicting future chloride contamination rates from data collected in the first few years of service.

Both carbonation and chloride diffusion rates are functions of temperature. As seasonal and daily variations are rapid compared with diffusion of chlorides or carbon dioxide, these are assumed to average out. The environments in northerly and southerly latitudes change in temperature, relative humidity and exposure condition so separating out temperature effects is difficult.

8.3 RATE OF DEPASSIVATION (ACTIVATION)

We have treated the problem as one dimensional so far, considering the time to depassivation at one particular location. Carbonation depths, chloride profiles and rebar depths are not uniform so the spatial distribution of depassivation or initiation must be included in the calculation unless the ranges are small or the time from depassivation to damage is large. We know that all the concrete cover will not spall off at once so there must be a distribution of depassivation times and of time from depassivation to spalling. We must have realistic estimates of the time from the first spall to end of functional service life.

By looking at the distribution of diffusion coefficients and the cover distribution it should be possible to calculate T_0 for the first 1%, 10%, 20%, etc. of the structure. It may also be important to differentiate between different locations due to variations in exposure. This will include moving up a column from the sea level, areas of salt water run off on substructures, zones facing salt spray, etc.

8.4 DETERIORATION AND CORROSION RATES, T_1

The diffusion models work reasonably well for predicting the initiation time. The chloride profile and the carbonation depth can be measured in the laboratory and in the field. However, it is far more difficult to look at the next step in our model. Corrosion rate measurements are now being taken in the field with linear polarization instruments and empirical estimates have been made with different instruments for the time to spalling.

The following broad criteria for corrosion have been developed from field and laboratory investigations with the sensor controlled guard ring device (see Section 4.11):

$$\text{Passive condition: } I_{corr} < 0.1 \ \mu A \ cm^{-2}$$

$$\text{Low to moderate corrosion: } I_{corr} \ 0.1 \text{ to } 0.5 \ \mu A \ cm^{-2}$$

$$\text{Moderate to high corrosion: } I_{corr} \ 0.5 \text{ to } 1 \ \mu A \ cm^{-2}$$

$$\text{High corrosion rate: } I_{corr} > 1 \ \mu A \ cm^{-2}$$

The device without sensor control has the following recommended interpretation (Clear 1989):

$$\text{No corrosion expected: } I_{corr} < 0.2 \ \mu A \ cm^{-2}$$

$$\text{Corrosion possible in 10 to 15 years: } I_{corr} \ 0.2 \text{ to } 1.0 \ \mu A \ cm^{-2}$$

$$\text{Corrosion expected in 2 to 10 years: } I_{corr} \ 1.0 \text{ to } 10 \ \mu A \ cm^{-2}$$

$$\text{Corrosion expected in 2 years or less: } I_{corr} > 10 \ \mu A \ cm^{-2}$$

The relative sensitivities of the two instruments are discussed in Section 4.11 on corrosion rate measurement. These measurements are affected by temperature and RH, so the conditions of measurement will affect the interpretation of the limits defined above. The measurements themselves should be considered accurate to within a factor of two (Feliú et al., 1995).

For carbonation it seems that the rate of corrosion falls rapidly as the relative humidity in the pores drops below 75%, and rises rapidly to an R.H. of 95% (Tuutti, 1982). For any sort of corrosion, there is also approximately a factor of 5–10 reduction in corrosion rate with 10°C reduction in temperature. There is obviously a slowing to zero when pore water freezes, although this is somewhat below 0°C. The calculation of corrosion rates is therefore critically dependent upon having measurements at different times of year and under different (representative) conditions to derive a realistic typical or average corrosion rate.

The corrosion currents given above can be directly equated to section loss by Faraday's Law of electrochemical equivalence, where 1 $\mu A \ cm^{-2}$ = 11.5 μm section loss per year.

Various efforts have been made to estimate the amount of corrosion that will cause spalling. It has been shown that cracking is induced by less than 0.1 mm of steel section loss, but in some cases far less than 0.1 mm has been needed. This is a function of the way that the oxide is distributed (i.e. how efficiently it stresses the concrete), the ability of the concrete to accommodate the stress (by creep, plastic or elastic deformation) and the geometry of rebar distribution that may encourage crack propagation by concentrating stresses, etc. e.g. in a closely spaced series of bars near the surface, or at a corner where there is less confinement of the concrete to restrain cracking.

What we refer to as 'rust' is a complex mixture of oxides, hydroxides and hydrated oxides of iron having a volume ranging from twice to about six times that of the iron consumed to produce it. This assumes 100% density of the product. Any porosity will increase the volume further.

If we assume a volume increase of three on average (or four allowing for the 1:1 replacement of the consumed steel) then the corrosion rates above translate as follows:

0.1 $\mu A\ cm^{-2} \equiv 1.1\ \mu m\ yr^{-1}$ section loss $\equiv 3\ \mu m\ yr^{-1}$ rust growth

1.0 $\mu A\ cm^{-2} \equiv 11.5\ \mu m\ yr^{-1}$ section loss $\equiv 34\ \mu m\ yr^{-1}$ rust growth

5.0 $\mu A\ cm^{-2} \equiv 57.5\ \mu m\ yr^{-1}$ section loss $\equiv 173\ \mu m\ yr^{-1}$ rust growth

10.0 $\mu A\ cm^{-2} \equiv 115\ \mu m\ yr^{-1}$ section loss $\equiv 345\ \mu m\ yr^{-1}$ rust growth

From the corrosion rate measurements it would appear that about 10 μm section loss or 30 μm rust growth is sufficient to cause cracking.

8.4 The Clear/Stratfull empirical calculation

Stratfull developed an empirical equation to determine the time to first distress of reinforced concrete in sea water with a known, constant chloride content. Clear (1976) modified this to be used for atmospherically exposed structures:

$$T = [(0.052\ d^{1.22}\ t^{0.21})/(Z^{0.24}P)]^{0.83} \tag{8.4}$$

where T is the time to first cracking (years); d is the depth of cover in millimetres or depth of cover minus depth of near surface chloride measurement (Z); t is the age at which Z was measured (years); Z is the surface (or near surface) chloride concentration (per cent by weight of concrete); and P is the water/cement ratio.

The difficulty with using such a model is knowing when it is not valid as it is empirically based. Several other variations on this equation also exist (Purvis et al., 1992). It is difficult to know which is the most appropriate version to use.

8.5 CORROSION WITHOUT SPALLING

A different situation exists when high levels of water saturation and low levels of oxygen lead to the 'black rust' that is deposited in the concrete without exerting stresses. In this case the corrosion rate measurement can be taken as showing a section loss that will eventually lead to failure. However, it is very difficult to extrapolate an instantaneous measurement of corrosion rate to a total section loss measurement. If we can make a series of corrosion rate measurements at different locations and we can come up with a compensation for variations in relative humidity and temperature we can estimate the average corrosion rate. We also have to estimate the original time to corrosion, and assume that the corrosion rate has either been constant or increased in some sensible manner (say linear or logarithmical) to the present condition.

8.6 PITTING CORROSION

The problem of pitting is discussed under corrosion rate measurement. If an investigation reveals pits then we can assume the corrosion rate is about five times that measured with an accurate linear polarization device.

8.7 CRACKING AND SPALLING RATES, CONDITION INDICES AND END OF FUNCTIONAL SERVICE LIFE

We can predict the initiation time, and the time to cracking and spalling and give a distribution of that time to show how corrosion damage will spread with time and location across our structure. In order to decide at what point it must be repaired we must define an 'end of service life condition'. This is not an easy definition. It is often subjective and will vary from structure to structure. It will also vary from engineer to engineer. In a SHRP study of service life estimates where 60 North American highway engineers were questioned about their bridges it was found that bridge decks had reached the end of their functional service life when 5 to 14% of the deck was spalled, delaminated or patched with temporary asphalt patches. For substructures it was found that about 4% cracking, spalling and delamination represented end of functional service life (Weyers et al., 1994).

In a separate SHRP study (Purvis et al., 1994) a condition index S was derived:

$$S = [Cl + 2.5(Delam) + 7.5(Spall)]/8.5 \qquad (8.5)$$

where Cl is the percentage of bars with chloride above the corrosion threshold, $Delam$ is the percentage of surface with hidden delaminations and $Spall$ is the percentage of surface with visible deterioration.

No attempt was made to define an end of service life as this was used as part of a lifecycle cost analysis study where repair or rehabilitation was conducted according to the minimum cost criterion. It was suggested that the condition index should never exceed 45%.

Some engineers will define the end of functional service life as the time to first spall as this is the point at which remedial action will be undertaken. From a safety point of view this may be relevant but it may not be cost effective depending upon the importance of the appearance of the structure and the feasibility of containing and controlling concrete debris.

The link between cracking and spalling is difficult to define. However, it may be possible to look at the cracking rate and suggest that if 5% of bars have cracks then this will lead to 1% spalling. The condition index equation 8.5 suggests a 3:1 ratio of cracking to spalling. The SHRP researchers suggested that the percentage delamination is four times the spalling percentage for most concrete decks, eight times for decks with $\geqslant 25$ mm concrete overlays and for most substructures and 16 times for substructures with $\geqslant 25$ mm concrete jackets or shotcrete.

The authors also suggested that chloride contamination (percentage above the corrosion threshold) increases linearly from 0 at condition index 0 to 100% at condition index 20.

In the SHRP study the condition index was calculated as a function of time by fitting it to a curve of the form:

$$S_t = 100/[1 + A \exp(-Bt)] \qquad (8.6)$$

This is an S-shaped curve that approximates to the two straight lines for T_0 and T_1 in Figure 8.1. The curve could therefore be calculated if there are two sets of data. These can be derived from two sets of measurements of cracking, spalling and chloride content separated by several years, or taking one set of measurements at the present time t_p and back calculating data for an earlier time t_e. This may be to the time of the first spall or a back calculation of the time to depassivation from the chloride profiles (approximately T_0). We can therefore derive values for A and B. These can be used to project forward the delamination rate and show how costs will escalate if work is deferred or how repair quantities will increase between the survey and the start of patch repair work.

8.8 SUMMARY OF METHODOLOGY TO DETERMINE SERVICE LIFE

When presented with a corroding structure we can determine its condition by measuring the chloride profiles, carbonation depths and cover depths. From this we can calculate diffusion rates of the carbonation front or the chloride threshold and estimate the initiation time T_0. Both an average and a distribution of T_0 values can be derived.

By measuring corrosion rates over a period of time we can estimate the time to cracking and knowing the distribution of corrosion rates, cover depths, etc. a cracking rate can be established by adding the time to cracking to the initiation time. An empirical condition curve can be calculated and the time taken to reach an unacceptable level can be determined.

REFERENCES

Clear, K.C. (1976) *Time-to-Corrosion of Reinforcing Steel in Concrete Slabs*, FHWA-RD-76-70, Washington, DC.

Clear, K.C. (1989) *Measuring Rate of Corrosion of Steel in Field Concrete Structures*, Transportation Research Record, Transportation Research Board, Washington, DC.

Feliú, S., Gonzalez, J.A. and Andrade, C. (1995) 'Electrochemical methods for on-site determination of corrosion rates of rebars', in *Symposium on Techniques to Assess the Corrosion Activity of Steel in Reinforced Concrete Structures*, December 1994, American Society for Testing and Materials.

Germann Instruments (1994) *A/S RCT Profile Grinder Mark II Instruction and Maintenance Manual*, Germann Instruments, Denmark.

Miller, J. (1995) Personal communication.

Parrott, L.J. (1987) *A Review of Carbonation in Reinforced Concrete*, A C and CA Report for Building Research Establishment, Watford, UK.

Parrott, L. and Chen Zhang Hong (1991) 'Some factors influencing air permeation measurements in cover concrete', *Materials and Structures*, 24, 403–8.

Parrott, L.J. (1994a) 'Moisture conditioning and transport properties of concrete test specimens', *Materials and Structures*, 27, 460–8.

Parrott, L.J. (1994b) *Carbonation-Induced Corrosion*, Paper presented at the Institute of Concrete Technology Meeting, Reading, 8 November, Geological Society, London.

Poulsen, E. (1990) *The Chloride Diffusion Characteristics of Concrete: Approximate Determination by Linear Regression Analysis*, Nordic Concrete Research No. 1, Nordic Concrete Federation.

Purvis, R.L., Graber, D.R., Clear, K.C. and Markow, M.J. (1992) *A Literature Review of Time-Deterioration Prediction Techniques*, Strategic Highway Research Program, National Research Council, SHRP-C/UFR-92-613.

Purvis, R.L., Babaei, K., Clear, K.C. and Markow, M.J. (1994) *Life-Cycle Cost Analysis for Protection and Rehabilitation of Concrete Bridges Relative to Reinforce-*

ment Corrosion, Strategic Highway Research Program, National Research Council, SHRP-S-377.

Tuutti, K. (1982) *Corrosion of Steel in Concrete*, CBI Swedish Cement and Concrete Institute, Stockholm.

Weyers, R.E., Fitch, M.G., Larsen, E.P., Al-Qadi, I.L., Chamberlin, W.P. and Hoffman, P.C. (1994) *Concrete Bridge Protection and Rehabilitation: Chemical and Physical Techniques – Service Life Estimates*, Strategic Highway Research Program, SHRP-S-668, National Research Council, Washington, DC.

9
Building for durability

9.1 COVER, CONCRETE AND DESIGN

Figure 9.1 shows what can happen in cases of inadequate design and maintenance of structure subject to chloride attack. The bridge is in Ohio and has a sign underneath saying 'Beware of falling rocks'. The car park was also in the Midwestern USA and collapsed overnight without injury or loss of life but with considerable expense to the owner.

The first requirement for maximum durability against corrosion is low permeability concrete, with a high cement content, a minimal chloride content and good cover to the reinforcing steel. Good concrete and good cover should be designed in and enforced during construction regardless of other requirements and additional protection. Good cover and low water/cement ratios will increase the time needed for chlorides and carbonation to reach the reinforcing steel.

Additives and cement replacement materials such as pulverized fuel ash (fly ash), ground granulated blast furnace slag, silica fume (micro silica) and other materials can reduce the pore size and block pores enhancing durability further. However, proper curing may be vital to get the required performance with these blended cements.

The British Standard on design of concrete bridges, BS5400, Part 4, gives guidance on durability and environment, particularly concerning chloride environments. The nominal cover requirements for different corrosion environments are given for different concrete grades. However, the document *Design for Durability* (Department of Transport, 1994) increases these requirements by 10 mm. This specified increase may be due to the difference between nominal and actual cover. To ensure that 95% of the bars achieve an actual cover of 75 mm it may be necessary to specify a nominal cover of 85 mm. The document also notes that such high cover can lead to shrinkage cracking which may need to be controlled by adding fibres to the concrete.

200   *Building for durability*

Figure 9.1(a) and **(b)** The consequences of poor design for durability and the 'do nothing' maintenance option. Courtesy Joe Lehman and Jack Bennett.

When designing a structure, detailing and awareness of the climate and microclimate can avoid long-term problems. Installing waterproofing membranes on decks (in bridges and car parks), applying coatings or cladding to areas subjected to spray from vehicles or sea water will all delay the ingress of moisture and chlorides.

The designer should also make it as easy as possible to assess and repair the structure. One problem with post-tensioning is that there is no easy way to assess the condition of the tendons in the ducts. Even if we find corrosion how do we repair it? In the UK a committee has been set up to answer the questions of design for durability, assessment and repair of post-tensioned, ducted concrete bridges. Their interim report has been published (Concrete Society, 1995). It deals with design, construction and quality systems for building new durable bonded post-tensioned structures. The problem of assessing the hundreds of existing structures that may be suffering from corroded tendons is now being undertaken by the Highways Agency.

The UK DoT *Design for Durability* Advice Note and Standard (1994) give clear guidance on minimizing the amount of reinforcement available for corrosion, maximizing access, improving drainage and minimizing runoff of salt laden water. Use of drips and proper guttering to control runoff is also discussed.

For structures with a very long lifetime requirement such as tunnels and other major civil engineering structures it is possible to carry out calculations to predict the likely time to initiation of corrosion (see Chapter 8). The chloride concentration at the surface can be predicted or estimated from structures in similar conditions. If special concrete mixes are being designed then the diffusion constant can be calculated according to the equations in Chapter 8 and the time for the corrosion threshold to reach the steel can be calculated. Calculating the consequences of reduced cover is also possible.

Other methods are available for enhancing durability by protecting the concrete surface, adding further protection to the concrete and protecting the steel. These are all discussed below. Another interesting issue is the installation of cathodic protection on new structures. This is also discussed at the end of the chapter.

9.2 FUSION BONDED EPOXY COATED REBARS

The protection system of choice on highway bridges in North America for aggressive chloride conditions is fusion bonded epoxy coated reinforcement (FBECR). In the 18 years since its first use on a bridge deck, epoxy coated reinforcement has been used in more than 100 000 structures in the USA and Canada. This equates to over 2 million tonnes of epoxy coated rebar.

There are several reasons for FBECR becoming the protective system of choice in the USA and Canada for reinforcing steel exposed to chloride attack. One is the reluctance to use waterproof membranes on bridge decks as they are difficult to install properly and to monitor both for correct installation and for performance after installation. The preference for a very low maintenance bridge deck led most state DOTs and the Federal Highway Administration (FHWA) to look for alternative protective systems on all bridges exposed to chlorides from the sea or from deicing salt.

In 1982 a panel consisting of David Manning (Ontario DOT), Ed Escalante (National Bureau of Standards) and David Whiting (Construction Technology Labs/Portland Cement Association) reviewed the performance of galvanized reinforcing steel. They recommended that it was not likely to give the long-term performance required (more than 40 years). This lead to the US recommendation of FBECR over galvanized rebar.

Although the corrosion inhibitor admixture calcium nitrite has been investigated and was considered acceptable by the FHWA its performance in tests was not as good as FBECR. The lack of competitive tender due to the patent being owned by a single manufacturer may also make states reluctant to specify it. Problems of flash setting and freeze–thaw damage have also been of concern when using nitrite inhibitors if the mix design is not adequately thought out. However, after their problems with FBECR the Florida DoT is looking at micro silica concrete and calcium nitrite admixtures to achieve corrosion resistance along the coast line.

By a process of elimination rather than design, this has led to the preeminence of FBECR as the main protection of reinforcing steel from chloride attack in North American highway bridges, car parks and marine structures. In some Canadian provinces both waterproof membranes and FBECR are used.

In Europe comparatively little FBECR is used, except in dowel bars for concrete pavements, where they have performed well for several years. They are also used for electrical isolation in tram and light railway system and for power transmission systems.

Compared with North America very few plants exist in Europe for manufacturing FBECR. The extensive use of waterproof membranes had minimized the problem of potholes on bridge decks. However, massive deck and joint repairs have been necessary over the last few years on many elevated sections of UK motorways due to chloride penetration through and round the ends of membranes. The first major European project to specify FBECR was the Great Belt Tunnel. This application is described in the next section.

9.2.1 How does epoxy coating work?

Epoxy coatings are applied to reinforcing steel in a factory process. The bar is grit blasted clean, it may then be pretreated and it is then heated in an induction furnace and passed into a coating unit that sprays a fine epoxy powder at the bar. The powder fuses onto the hot bar, is cured and quenched before passing out of the process (Figure 9.2a).

The important property of the fusion bonded epoxy coating is that it is a dielectric, and charged species such as the chloride ion cannot pass through it. It also has excellent adhesion properties to the steel and is not easily undercut by corrosion at defects. It must also be flexible to allow the straight bars to be bent during fabrication on a special mandrill to protect the coating from damage (Figure 9.2b).

An alternative method of applying and using epoxy coated rebar has been used on the Great Belt Bridge/Tunnel lining segments in Denmark and also the tunnel lining for the St Claire tunnel between the USA and Canada. In these projects the rebar cages are fabricated and

Figure 9.2(a) An epoxy coating plant in action. Acknowledgements to Allied Bar Coaters, Cardiff, Wales.

Figure 9.2(b) Bar bending in the factory. Acknowledgements to Allied Bar Coaters, Cardiff, Wales.

the bars welded together. Instead of the coating being applied in a linear production line process the prepared cages are dipped in a fluidized bed of epoxy powder (Figure 9.3). Immediately after coating the cages can then be cast into the concrete with least risk of damage to the coating.

Defects are inevitable in any coating process. These come from several sources. There are pinholes in the coating after application because reinforcing steel is not smooth and has deformations on it. Small amounts of damage are done to the coating during handling and fabrication (bending) in the manufacturing plant. Some damage is inevitable during transportation to site, and during the assembly of the reinforcement cages. Finally, the act of pouring wet concrete over the reinforcement and then vibrating it may cause further damage to the coating. The aim of all current standards on epoxy coated reinforcement is to keep the number and size of defects low and to repair them when possible. By keeping the number low, the size small and the separation high with a coating that minimizes undercutting corrosion there should be effective separation of anodes and cathodes (see Chapter 2). This increases the resistance in the corrosion cell and decreases the corrosion rate (Figure 9.4).

Epoxy coated rebars will corrode eventually if the chloride level

Figure 9.3 A prewelding rebar cage emerging from the dipping tank on the Storebelt (Great Belt) tunnel manufacturing line. Acknowledgements to Mott MacDonald.

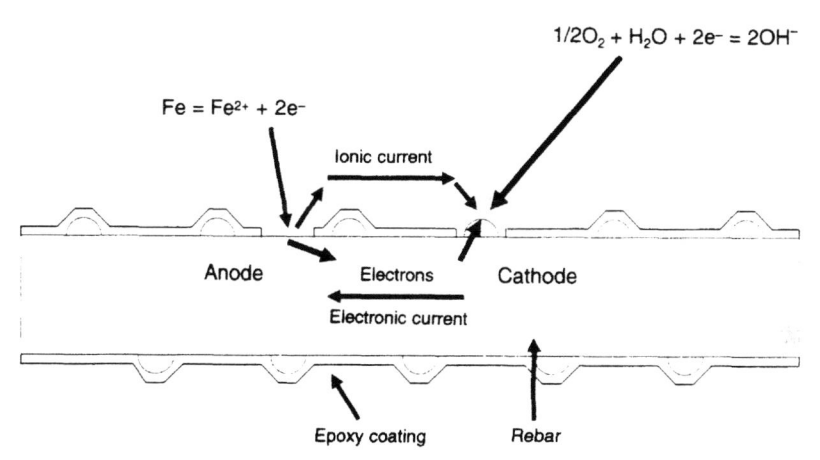

Figure 9.4 Schematic of corrosion between anodes and cathodes at defects on an epoxy coated rebar.

builds up to a high enough level at the rebar surface and there is moisture and oxygen to fuel the corrosion reaction. However, they should take longer to start corroding than uncoated reinforcing steel and then should corrode more slowly than uncoated rebars.

9.2.2 Problems with epoxy coating

The problems found in the Florida Keys were first identified in the late 1980s. The Keys bridges were built in the late 1970s with FBECR. Routine inspection revealed localized corrosion in 1986 on V pier struts. Further corrosion was identified on a total of five bridges in subsequent inspections. The author visited the Keys in 1989 and saw extensive delamination in the tidal and splash zones with severe corrosion of the rebar exposed when delaminated concrete was removed with a hammer.

The following points should be noted when reviewing the Keys bridge corrosion problems.

1. Corrosion was localized to the splash and tidal zones.
2. There was no clear relationship between low cover and corrosion.
3. Although chloride levels were high, they were no higher than in non-corroding bridges in the Keys and elsewhere in Florida.
4. The other bridges in the Keys and 10 bridges elsewhere in Florida containing FBECR have been examined and are not yet showing corrosion problems. At least 20 bridges have been examined in detail.

It is now believed that poor handling practice with excessive exposure of coated bars to ultraviolet light, salt and mechanical damage had a major role to play in the failures.

There is, however, a fundamental concern about the adhesion of the coating to the steel. FBECR is corrosion resistant because the epoxy coating is a dielectric film that stops charged ions (chloride) from passing through it. The other option of the ion is to go round it, e.g. at defects or holidays in the coating. At these locations the ion should not be able to get under the coating to form a pit (see Chapter 2). A new, good quality FBECR shows excellent adhesion of the coating to the steel and cannot be peeled off.

However, many bars in the Keys bridges, both those showing corrosion and those still in good condition, had poor adhesion and the coating could be peeled off the steel. There were also problems on new bars. An immediate programme of tightening application practice was undertaken. One major problem for adhesion is any deposits left on the surface after grit blasting and before coating. Salt is a particular problem as bars are often exposed to road and sea salt during transport to the coating plant and once in the grit it can redeposit on the bar surface as the grit is recycled.

Even with good initial adhesion there was a loss of adhesion for concrete specimens exposed to regular or continuous wetting. Adhesion testing on the deformed surface of a rebar is difficult to perform quantitatively and there is some disagreement over whether the loss of adhesion when wet is permanent or fully recovered if the bar dries out. Even if it is temporary, the bar is more susceptible to corrosion and under-film attack at defects while wet and once corrosion has started it will not recover adhesion.

Major research programmes are now underway in the USA and in other countries to find out whether the loss of adhesion is a fundamental problem in the durability of FBECR. If it allows extensive under film corrosion such as that seen on the Florida Keys bridges then FBECR may not be a cost-effective solution to enhanced durability. The latest results are given in Clear et al. (1995).

All studies have shown that FBECR outperforms bare steel and galvanized steel in the laboratory and in the field. However, its detractors have suggested that the improved durability of field structures is associated with the reduced water:cement ratio of the concrete and increased cover that were implemented on all structures when using FBECR. For the foreseeable future it is likely that most North American highway departments will continue to specify FBECR and most European departments will not. Further tightening of specifications to avoid damage and to get the highest possible quality of coatings and coating adhesion have been implemented and may be improved further over the next few years. There is also research on how much damage occurs to the coating between leaving the coating plant and casting into concrete.

The position is changing rapidly and alternative corrosion protection methods such as the use of micro silica and calcium nitrite are being considered by Florida DOT and may spread to other major North American users of FBECR.

It is likely that the prefabricated fusion dipped cage approach will have long-term durability and will be applied on special projects. This was first used on the Great Belt Bridge/Tunnel in Denmark as described above. It has also been applied to the St Claire Railroad Tunnel between the USA and Canada. As the owners are looking for an almost indefinite life out of the structure, concrete mix designs were developed to minimize chloride diffusion.

9.2.3 Advantages and disadvantages of fusion bonded epoxy coated rebars

There are certain disadvantages to using epoxy coated rebars. Their insulating properties mean that they inhibit the use of electrical and electrochemical techniques. Taking half cell potential measurements on epoxy coated rebar is very difficult. Since only small areas of the steel are exposed they are not necessarily in an alkaline environment as the pore water may not reach the steel, especially in areas of undercutting. It is also very difficult to interpret the measurements. For corrosion rate measurements with the linear polarization technique it is impossible to calculate the area of corrosion. For rehabilitation it is necessary to make electrical connections between all the rebars for cathodic protection or chloride removal.

Special handling techniques must be used at the fabrication plant, in transit to the site and during handling and storage on site. It is essential that damage to the coating is minimized.

There are problems with applying cathodic protection, chloride removal and realkalization to FBECR because the rebars are electrically isolated. Special care must be taken to connect each rebar together. This problem does not arise with the welded cages where each cage is continuous.

There are also problems with taking half cell or corrosion rate measurements. This is partly due to the electrical isolation of bars from each other as mentioned above, and partly because the steel is only exposed at small defects. The exposed steel at defects is not necessarily in an alkaline environment (either with or without chlorides) so interpretation of half cell potentials is difficult (see Section 4.7). For effective corrosion rate measurement the surface area must be known. If corrosion is only occurring at a few defects it may be impossible to determine the area of corrosion and therefore turn the corrosion current into an effective corrosion rate per unit surface area (see Section 4.11).

An important advantage of FBECR is that it is a comparatively low cost approach to corrosion protection. It is easily understood by site operatives and is now a routine technology in North America and some parts of the Middle East. If good quality rebars are cast into the concrete then there should be negligible maintenance requirements compared to waterproofing membranes or other surface coatings. There is no major impact for the bridge design other than an increase in lap lengths to take account of the reduced pull out strength of a bar with a deformable coating on it.

9.3 WATERPROOFING MEMBRANES

Waterproofing membranes are routinely applied to new bridge decks in the UK and much of Europe. Several DOTs in the Northern USA and Canada also apply them. Membranes come as two main types, a liquid that solidifies in place, and a sheet that is stuck to the concrete (see Section 5.3.3). The main problem with waterproof membranes is similar to that of epoxy coating rebars; they must be applied without defects. Defects can occur as blow holes, penetrations or mechanical damage to the liquid applied types, or cuts, tears, bad joins or perforations in the sheets. Either type of membrane can be damaged by the overlay of asphalt, either by its heat or mechanical damage from the aggregate particles.

Careful detailing is needed to avoid water and chlorides getting under the edges of the membrane. It is also essential to ensure that the deck and membrane drain properly, with adequate waterproofing around drains and with no ponding. The drainage system from the deck must not then dump salt laden water onto the substructure.

Many bridges in the UK (e.g. the Midland Links flyover and the Tees viaduct) have required extensive concrete repair to the bridge decks where membranes have failed to stop water and chloride ingress. The repair problems are worse at joints, where the membrane ends. The sealing of joints and making membranes continuous across them is a popular solution to this problem. Modern bridge design requires the minimizing of the number of joints in a bridge deck. There must be adequate detailing and proper attention to ensuring that the water does not accumulate and that membranes run into drains and over concrete edges where salt might accumulate.

Problems are also found at curbs, around drains and other places where the salt water can get under the membrane.

Waterproofing membranes have a life of about 15 years and then must be replaced. At that time it is good practice to survey for corrosion and chloride penetration after the membrane has been exposed. One problem with membranes is determining whether they even exist under the asphalt wearing course, or whether they are still working. It is sometimes possible to flood the deck with water and measure for electrical continuity between the surface and the reinforcing steel. This is difficult and will not find the defect. SHRP carried out research on using pulse velocity to define a condition rating for an asphalt overlayed membrane. This showed promise but needs further research before it can be developed into an accurate test.

The use of waterproofing membranes requires careful detailing of a structure at the design stage with good drainage and effective prevention of salt laden water getting under the membrane. This must be carried through to the construction site. A good finish must be established for the membrane to be applied successfully without debonding. Application of membranes requires a specialist to ensure there are no defects and that the membrane is adequately bonded to the concrete and sealed at the edges. The asphalt wearing course must be carefully placed to prevent damage to the membrane. The complete design of the structure must take into account the extra dead load and clearance required for the asphalt.

Some waterproofing membranes on car parks do not require asphalt concrete overlays. The extra wear on these membranes, especially from scuffing of tyres as cars manoeuvre in and out of parking spaces, means that they have shorter lifetimes and need more monitoring and maintenance.

9.3.1 Advantages and disadvantages of waterproofing membranes

If membranes fail, detecting the failure is difficult as they are overlaid with an asphalt wearing course. Leakage can occur through thinner slabs such as those on car parking structures, and traffic will degrade the asphalt overlay which will require replacement at regular intervals. There can be particular problems in car parks due to the manoeuvring of cars into parking spaces and round tight corners displacing the asphalt and damaging the membrane. Car park membranes often combine the waterproofing layer and the wearing course in one. Where a car park is built over offices or shopping malls leaking membranes can cause considerable problems.

Membranes have given good service on European bridges and in the New England states and other North American agencies where they are routinely used. They have reduced or eliminated the problems of potholing on bridge decks but at the price of a higher regular maintenance requirement compared with epoxy coated rebars in bridge decks.

Until recently membranes were not considered compatible with cathodic protection as the anode would have to be under the membrane to pass current, and the anode generates gases that must escape. Membranes and cathodic protection have been used in car parks on thin slabs with some success but these are still experimental. Chloride removal (or realkalization) could be done before the replacement of the membrane.

9.4 PENETRATING SEALERS

Several European DOTs specify the application of penetrating sealers to exposed concrete on bridges exposed to salt spray or run off. These are discussed in Section 5.3. They have very particular application requirements. In the US sealers are also used on bare concrete bridge decks. Their effectiveness on trafficked surfaces has not been fully investigated.

The advantage of silane type penetrating sealers is that they are inexpensive compared with cathodic protection, epoxy coated rebars or other preventative or active treatment methods. If penetration is good then there is no maintenance and a long lifetime (in theory) as the silane forms a hydrophobic layer inside the concrete pores. There is no impact on the design or performance of the structure. The disadvantages are the problem of ensuring adequate application of a colourless liquid, and knowing that it has penetrated the structure.

Applying cathodic protection or chloride removal through penetrating sealers if necessary is usually possible.

9.5 GALVANIZED REBAR

As stated in Section 9.2, US research has generally shown that galvanizing is an inferior option to fusion bonded epoxy coated reinforcement when faced with chloride attack. Galvanizing has one principal advantage over FBECR; it will accept damage in handling as the coating corrodes sacrificially and defects are not as important as for FBECR. Galvanizing is used in many other countries to good effect. The FHWA memorandum suggests a 15-year life for galvanized rebar in good quality concrete. Their conclusions are supported by Andrade *et al.* (1994) with respect to chloride attack. Galvanized bars may be more effective against carbonation.

Galvanized rebar is used successfully in structures where carbonation is a risk such as cladding panels. Galvanizing can easily be carried out in most countries although the quality and composition of the coating can affect its durability. It suffers from fewer problems when handled roughly because the coating is sacrificial and protects bare areas.

Accelerated depletion of the galvanizing can occur if galvanized rebars are mixed with ungalvanized bars. If they are being used in the same structure (for instance on a bridge deck but not the substructure or vice versa), then care should be taken to ensure complete electrical isolation of the galvanized and ungalvanized bars.

It may be very dangerous to apply electrochemical treatments to galvanized reinforcing steel. Very severe pitting can result. NCT, the patent holders on the realkalization and desalination techniques, do not recommend their use on structures containing galvanized rebar (Miller, 1995).

9.6 STAINLESS STEEL REINFORCEMENT

Stainless steel rebar has been applied in special circumstances but it is a very expensive option. A duplex steel with a mild steel core and stainless outer 1–2 mm has also been offered, which is cheaper, but still relatively expensive. Stainless steels can be susceptible to pitting attack, so the correct grade of stainless steel must be used.

The problem of mixing metals applies to stainless steel with mild steel as with galvanized and ungalvanized steel in the section above. The galvanic coupling of mild steel and stainless steel will accelerate corrosion of the mild steel. This problem has been observed on balconies and facades (Miller, 1995).

9.7 CORROSION INHIBITORS

Calcium nitrite is the principal corrosion inhibitor available to stop corrosion that is compatible with concrete in the casting process. As stated in Section 9.2, it is accepted by FHWA as an alternative to FBECR for protection against chloride induced corrosion. FHWA research shows that if sufficient nitrite is added to the concrete mix to ensure a chloride to nitrite ratio of less than 1.0 at rebar depth, then the nitrite will prevent corrosion. Obviously this is feasible in marine conditions where the chloride level is known (and assuming no concentration effects) but may be more difficult in other situations.

After their unfortunate problems with epoxy coated reinforcing steel, Florida DoT is considering the use of calcium nitrite inhibitor against corrosion. This has the advantage that there is continuity of the reinforcing cage so electrochemical methods can be used to measure for corrosion (half cells and corrosion rate) and to treat for corrosion (cathodic protection and chloride removal). Its major drawback is that with a good quality, dense concrete (Florida DoT now specifies micro silica in the concrete) and good cover, the inhibitor should not be needed for at least 20 years. Will it still be there and will it still be effective 20 years after being put in the concrete mix?

The advantage of calcium nitrite is that it can be added to the mix and has no serious effects on the design, construction and performance of the structure other than its effect as a set accelerator. Mix design may require adjusting to include a retarder. Its disadvantage is that there must be enough to stop corrosion and it is consumed with its exposure to chlorides. It is therefore important to calculate the chloride exposure for the life of the structure and add sufficient inhibitor. It does not inhibit the application of cathodic protection or chloride extraction in later life of the structure if necessary.

9.8 INSTALLING CATHODIC PROTECTION IN NEW STRUCTURES

There are two reasons for installing cathodic protection from new. One is to energize it initially as a 'cathodic prevention system'. The other reason is to have it there for use at a future date.

The first approach has been pioneered on Italian Autostrada bridges in the mountainous areas of northern Italy (Pedeferri, 1992). Previous construction experience had shown that it was very difficult to prevent chloride ingress. Therefore cathodic protection anodes were either built into the deck or applied as part of the construction process. The bridges were post-tensioned segmental construction with tendons in grouted ducts.

Cathodic protection was applied at a very low level to avoid any risk of hydrogen evolution. As chlorides had not entered the structure the aim was to apply a small level of current to polarize the steel and effectively repel the chlorides. These systems were installed in the late 1980s and early 1990s. If properly maintained they should give excellent life as the anodes will be used at a very low rating. Another case of installation from new was a salt silo or chamber with a reinforced concrete floor. This was known to be at high risk from chloride penetration. Further examples are the rebuilt underground car park of the World Trade Center in New York, and a palace in the United Arab Emirates (Funahashi, 1995).

The installation of anodes and rebar connections during construction for later use has been discussed on a number of structures. While positive efforts to ensure rebar continuity at the construction stage are to be welcomed, it is generally considered that anode design and types will change over the next decade so improved systems may be available by the time that protection is required. If corrosion is local then the application of small, local anodes at the time of corrosion may be more cost-effective than designing in anodes to protect the whole of the structure. Experience has also shown that connection leads get damaged and destroyed or may not be in the correct location for future use. It is generally better to design the anode system to do the job when required rather than 10 or 20 years early.

9.9 DURABLE BUILDINGS

Most of this chapter has discussed bridges, with peripheral comments about other civil engineering structures. This is partly because most of the emphasis in the field of durability has been problems with bridges caused by chloride attack. Many of the discussions are relevant to all reinforced concrete structures. However, buildings are constructed to withstand less severe environments and conditions than bridges. There are also lower life expectations on buildings which may become obsolete before serious deterioration occurs.

The trend towards covering up concrete with coatings and cladding increases the durability of the concrete by retarding chloride and carbon dioxide ingress. Buildings rarely use FBECR, micro silica, sealers or inhibitors, but they rarely suffer from severe chloride exposure.

If sufficient attention is paid to ensuring the specified cover and water cement ratio, most buildings in relatively benign environments will perform well throughout their normal lifetimes. In aggressive environments such as many countries in the Middle East or where

214   *Building for durability*

exposed to sea salt spray, then designers should consider the durability codes for bridges rather than those for buildings.

Adequate maintenance and maintainability are also important. If leaks develop on the flat roof of a building exposed to the sea then chlorides can concentrate and cause damage. If drainage is well designed and maintain then problems will be minimized.

9.10 CONCLUSIONS

There are several methods of ensuring that reinforced concrete structures do not suffer from corrosion problems. The most important is

Table 9.1 Protection of new structures

Technique	*Pros*	*Cons*
Cover and concrete quality	Should be cheapest. Products of good design and site QA	Standards and codes may still be inadequate
Blended cements	Well documented	Can change mix workability characteristics and performance
Sealers (substructures)	Comparatively inexpensive, can be applied selectively to areas at risk	Can be difficult to apply properly. Durability uncertain
Waterproofing membranes (bridge decks)	Well known technology, comparatively inexpensive	Requires maintenance. Damage occurs at ends, drains, kerbs, etc.
Epoxy coated rebar	Well known technology, very low maintenance	Problems of quality control in construction and of repair if failure occurs
Galvanized rebar	Cheap, easy to handle, good for carbonation	Poor for chlorides
Stainless steel bars	Very effective	Very expensive
Corrosion inhibitors	Calcium nitrite known to be effective. Should not interfere with later repairs	Limited experience over whole life of structures. Expensive additive to whole structure if only part is at risk
Cathodic protection	Known effectiveness	Initial cost plus maintenance required for life of structure

ensuring that the concrete quality is high, with a low permeability, and the cover is good. Additional protection can be achieved by keeping chlorides away by design, such as removing joints in decks where they can allow chlorides to leak on to substructures or applying coatings or cladding to facades of buildings exposed to sea salt spray. Further options are offered in the way of surface barriers, coatings, membranes and sealers. These can be used for a variety of applications to protect specific elements or specific areas at risk. Additional protection can be found in the form of blended cements with pozzolanic additives that can retard ingress, such as micro silica, PFA (fly ash), etc. The addition of corrosion inhibitors to the mix can be included in this category, although it bridges the distinction between a barrier to corrodents and a technique offering corrosion protection to the steel.

The above techniques were concerned with stopping chlorides and carbonation reaching the surface of the steel. The next category is making the steel corrosion resistant. These include epoxy coating, galvanizing and stainless steel bars. The problems with the former are performance. The problem with the latter is cost. Coatings such as epoxy or zinc impede our ability to assess and repair damage when it does occur. Finally there is cathodic protection that can be applied from new and is being used on several new structures to stop chlorides reaching the bar and to stop the bar from corroding if the corrodents reach it. Table 9.1 summarizes the techniques and their pros and cons.

REFERENCES

Andrade, C., Holst, J.D., Nurnberger, U., Whiteley, J.J. and Woodman, N. (1994) *Protection Systems for Reinforcement*, Prepared by Task Group VII/8 of Permanent Commission VII CEB.

Clear, K.C., Hartt, W.H., McIntyre, J. and Seung Kyoung Lee (1995) *Performance of Epoxy-Coated Reinforcing Steel in Highway Bridges*, National Cooperative Highway Program, Report 370, National Academy Press, Washington, DC.

Concrete Society (1995) *Interim Technical Report: Durable Bonded Post-Tensioned Concrete Bridges*, Interim Technical Report CS 111.

Department of Transport (1994) *Design for Durability*, Departmental Advice Note BA 57/94.

Funahashi, M. (1995) 'Cathodic protection systems for new RC structures', *Concrete International*, 17 (7), 28–31.

Miller, J. (1995) Personal communication.

Pedeferri, P. (1992) 'Cathodic protection of new concrete constructions', in *International Conference on Structural Improvements through Corrosion Protection of Reinforced Concrete*, Documentation E7190, IBC Technical Services, London, UK.

10
Future developments

Speculating about where the future might lead in the field of corrosion detection and rehabilitation is interesting. We are all aware of the revolution in electronics, particularly digital electronics, and information science in the past 20 years. We should recognize that materials science and technology has also gone through a major revolution bringing us new plastics, ceramics, chemicals, etc. All of this will affect the construction and repair industry over the next few decades.

Within the construction industry there will always be a major cost constraint when you are using thousands of tonnes of materials to build a single structure. Steel and concrete are very cheap compared with alternatives, and they usually do the job very well. The constraints of site practice will always require very different skills from those in the factory. No doubt we will continue to see more prefabricated construction to bring the disciplines of the controlled factory environment to the construction site.

Looking at the materials that make up or can be incorporated into concrete, the concrete industry will continue to develop new additives and cement replacement materials to improve concrete properties. A new range of additives may be essential if we are to respond to more pressure to use waste and recycled materials in construction rather than using high quality virgin aggregates and to increase the use of cement replacement materials. The performance of cement is under review particularly in Europe, with the rapid strength gain to 28 days and then very little change from there. This is different from the slow, steady strength gains of concrete 20 years ago. There are concerns that this is having detrimental effects on the ability of modern concretes to resist chloride and carbon dioxide ingress.

The use of corrosion inhibitors for steel in concrete is still in its infancy. We can expect to see new ideas and materials for both new construction and for rehabilitation over the next few years.

At the moment we can use 'bare' steel, epoxy coated steel, or galvanized steel for most applications. The number of alternative coatings may multiply over the next few years, and epoxy coatings may change to deal with the problems seen in the Florida Keys. Research on new coatings is currently underway in the USA (McDonald et al., 1995). The technique of prefabricating rebar cages, coating in a fluidized bed and then casting into precast units was developed for the Great Belt Bridge/Tunnel in Denmark. This approach may gain acceptance in more civil engineering construction work for more routine construction.

Alternatives to steel reinforcement are being experimented with, including glass reinforced plastic reinforcing bars and bridge decks with a single layer of reinforcing steel at the centre. More concrete containing fibres may be used.

The latest Concrete Society (1995) report on durable bonded post-tensioned concrete bridges will lead to improved design and construction of post-tensioned structures. However, improved repair and assessment methods for the existing structures are needed.

In the field of cathodic protection, anodes are being developed all the time with new coatings and other new materials. Research on sacrificial anode cathodic protection is also going on to find out whether these simpler, lower maintenance systems can be used effectively on bridge decks, substructures and elsewhere. The new electrochemical techniques of chloride migration and realkalization will continue to develop and grow. In some cases they will compete with cathodic protection, in other cases they will be applied where cathodic protection is inappropriate or less appropriate. The expanding numbers of companies using these techniques in an increasing number of countries will lead to further innovation in methods and materials of application.

There is an increasing acceptance of the value of applying coatings to concrete to increase durability. The major revolution at the moment is the requirement to reduce the amounts of volatile emissions and therefore there is a strong trend towards water based coatings. Concrete is a difficult material to coat because it needs to breathe. Penetrating sealers are a very popular solution to the problem of coatings to repel chlorides. However, the coating industry will no doubt continue to innovate and offer new alternatives.

Then there is the salt. Surprisingly, the costs of corrosion, massive as they are, still do not tilt the financial balance in favour of any of the alternative deicers or salt containing corrosion inhibitors except in a few very specialized applications (TRB, 1991). This is because salt is a very good deicer, it is very inexpensive and its environmental impact is known. All synthetic alternatives are less effective, more expensive and generally must be applied to a lot of highway to protect a few bridges. The road to a cost-effective alternative to deicing salt is going to be a long one. However, more effort is being spent on ensuring that salt is used as effectively as possible to minimize the amount used and to make its application as cost-effective as possible.

On the design side, efforts will continue to be made to 'design out' corrosion. There will always be a balance between cost and durability. The use of less reinforcing steel increases the size and weight of components. Increased cover increases the risk of surface cracks, which can be counter-productive. Increased site supervision to ensure that proper cover is achieved will increase construction costs. However, it is likely that less reinforcing steel will be used in situations where it is vulnerable to corrosion and improved site practice and quality systems will improve durability.

As we improve our ability to model corrosion and predict durability it should be possible to give proper engineering judgements on initial cost versus maintenance and repair costs. Is it worth paying for permeable formwork, pozzolanic additives and other measures that can improve the concrete quality? What is the cost-effective balance between accurately controlled cover or various additives in terms of decreased lifecycle cost? The separation of the capital cost of construction from the maintenance cost requires new attitudes towards accounting in the private and the public sector. Lip service is paid to these ideas but with tight budgets it is always easier to take the lowest bid. However, at least in maintenance contracts, there is a move toward better contracting procedures such as the 'two envelope' bid system where the price envelope is only opened after the technical approach has been approved, so that only the best bids are considered.

Improvements in our ability to survey and monitor structures will continue. With the availability of modern digital processing and communications systems, permanent corrosion monitoring is increasingly economically feasible and attractive, especially on prestige civil engineering structures. Corrosion rate measurement is in its infancy so we can expect new methods and improved information about corrosion. As well as understanding the effect of the environment on structures, corrosion monitoring will help us to understand the 'microclimate' that leads to corrosion. We will also be able to study the effectiveness of repair systems in different environments and applied to different structures.

One problem faced by the innovator in the marketplace is the time needed for a return on investment for developing a new product compared with the time the construction industry takes to accept a product. A ten year time scale from prototype to acceptance is too long for most investors. However, that is typically the time that the civil engineering industry takes to accept new technology. New performance standards may help the innovator by giving better, fixed 'goal posts' to aim at. However, the industry still has to accept and try new products that look promising or the flow of new products will dry up and the innovators will move into more rewarding fields. If that means we will have some failures, then perhaps that is the price to be paid for the successes.

We will always need to build structures in corrosive environments. We will therefore need new techniques to deal with the problems that result. We will also continue to refine the techniques that we have. The construction and repair industry will continue to need engineers and corrosion specialists with knowledge and experience of dealing with the problems as they are the most important resource in the fight against corrosion. Good undergraduate and postgraduate courses on corrosion are needed as well as research projects to train young engineers in the theory and practice of corrosion of steel in concrete.

REFERENCES

Concrete Society (1995) *Durable Bonded Post-Tensioned Concrete Bridges*, Interim Technical Report, Slough, UK.

McDonald, D.B., Sherman, M.R. and Pfeifer, D.W. (1995) *The Performance of Bendable and Nonbendable Coatings for Reinforcing Bars in Solution and Cathodic Protection Tests*, FHWA Report FHWA-RD-94-103, Federal Highway Administration, McLean, VA.

Transportation Research Board (1991) *Highway Deicing: Comparing Salt and Calcium Magnesium Acetate*, TRB Special Report 235, Transportation Research Boards, National Research Council, Washington, DC.

Appendix A

Bodies involved in corrosion and repair of reinforced concrete

American Concrete Institute (ACI)
PO Box 19150 Redford Station
Detroit, MI 48219, USA
Tel. + 1 313 532 2680
Fax + 1 313 538 0655

American Society for Testing and
Materials (ASTM)
100 Barr Harbor Drive
West Conshohocken
PA 19428-2959, USA
Tel. + 1 610 832 9500
Fax + 1 610 832 9555

ASTM (Europe)
27–29 Knowl Piece
Wilbury Way
Hitchin
Hertfordshire SG4 0SX, UK
Tel. + 44 1462 437933
Fax + 44 1462 433678

Concrete Repair Association
PO Box 111
Aldershot
Hampshire GU11 1YW, UK
Tel. + 44 1252 21302
Fax + 44 1252 333901

Concrete Society
3 Eatongate
Windsor Road
Slough SL1 2AJ, UK
Tel. + 44 1753 693313
Fax + 44 1753 692333

FERFA, Federation of Resin
Formulators and Applicators
1st Floor
241 High Street
Aldershot
Hampshire GU11 1TJ, UK
Tel. + 44 1252 342072
Fax + 44 1252 333901

Institute of Corrosion
PO Box 235
Leighton Buzzard
Bedfordshire LU7 7WB, UK
Tel + 44 1252 342072
Fax + 44 1252 333901

International Concrete Repair
Institute (ICRI)
1323 Shepard Drive, Suite D
Sterling, VA 20164-0116, USA
Tel. + 1 703 450 0116
Fax + 1 703 450 0119

Appendix A

NACE International
(Formerly the National
Association of Corrosion
Engineers)
PO Box 218340
Houston, TX 77218-8340, USA
Tel. + 1 713 492 0535
Fax + 1 713 492 8254

NACE International (Europe)
PO Box 47
Godalming
Surrey GU7 1TD, UK
Tel. + 44 1483 418299
Fax + 44 1483 418928

Scandinavian Society for
Corrosion Engineering
PO Box 13034
S-25013
Helsingborg, Sweden
Tel. + 46 42 16 22 78
Fax + 46 42 16 29 92

Society for the Cathodic Protection
of Reinforced Concrete (SCPRC)
Association House
235 Ash Road
Aldershot
Hampshire GU12 4DD, UK
Tel. + 44 1252 21302
Fax + 44 1252 333901

Transportation Research Board
2101 Constitution Avenue, NW
Washington, DC 20418, USA
Tel. + 1 202 334-2379
Fax + 1 202 334-2003

RILEM
Ecole Normal Supérieure
Pavillion des Jardins
61 avenue Du Pdt. Wilson
F-94235 Cachan Cedex, France
Tel. + 33 1 47 40 23 97
Fax + 33 1 47 40 01 13

Appendix B

Strategic Highway Research Program: published reports on concrete and structures (concrete and the corrosion of steel in atmospherically exposed reinforced concrete bridge components suffering from chloride induced corrosion)

Published reports are available from the Transportation Research Board (National Academy of Sciences) Publications Department, Box 289, Washington, DC 20055, USA.

The Strategic Highway Research Program (SHRP) was a five-year, $150 million programme, primarily aimed at addressing the major durability and maintenance issues on highways. It was paid for by the US state highway agencies and controlled by them.

The issues addressed were specification and durability of asphalt pavements (approximately $50 million), long-term pavement performance ($50 million), highway maintenance (about $10 million), winter maintenance ($10 million), concrete ($10 million) and structures ($10 million). SHRP started its first contracts in September 1987 and completed its research at the end of 1993.

The structures programme was entirely concerned with corrosion of reinforced concrete bridges suffering from salt induced corrosion. Its work on structures covered physical assessment, cathodic protection, electrochemical chloride removal, physical and chemical methods of rehabilitating bridge components and a methodology of bridge rehabilitation. There were also some extra smaller research contracts covering related issues. A complete list of the SHRP structures reports and those from the related concrete programme are included in this appendix and Appendix C.

CONCRETE AND STRUCTURES: PUBLISHED REPORTS

Concrete and Structures: Progress and Products Update. Describes the evolution of the concrete programme up to 1989. Objectives and expected products of the research are discussed, as well as their economic and technical significance. 62 pages. SHRP-C-300, $5

High Performance Concretes: An Annotated Bibliography 1974–1989. Over 800 references from the past 15 years are presented. 403 pages. SHRP-C-307, $5

Electrochemical Chloride Removal and Protection of Concrete Bridge Components. Discusses the feasibility of injecting synergistic inhibitors for protecting concrete bridge components. 47 pages. SHRP-S-310, $10

Handbook for the Identification of Alkali–Silica Reactivity in Highway Structures. Provides guidance for the field identification of alkali–silica reactivity (ASR) in Portland cement concrete structures such as highways and bridges. ASR development is assessed on two bases; the occurrence and disposition of cracking and displacement of concrete, and the presence of reaction products from ASR. Colour photographs. 48 pages. SHRP-C-315, $10

High Performance Concretes: A State-of-the-Art Report. Summarizes results of a literature review on the mechanical properties of concrete, with particular reference to the highway application of high performance concrete (HPC). Discusses the selection of materials and the manufacture of high performance concrete; the behaviour of plastic and hardened concrete; the behaviour of fibre-reinforced concrete; and the applications of high performance concrete. 250 pages. SHRP-C-317, $10

A Guide to Evaluating Thermal Effects in Concrete Pavements. Describes use of tables developed to help determine problems that result from early thermal effects in concrete. Parameters such as concrete temperature, air temperature, cement type and content affect the thermal behaviour of concrete. The tables help predict whether pavement temperature will become too high; whether temperature differences between the concrete slab or base and the air will result in early thermal cracking. 104 pages. SHRP-C-321, $10

Condition Evaluation of Concrete Bridges Relative to Reinforcement Corrosion. Vol. 1: *State of the Art of Existing Methods.* Reviews existing methods to detect damage caused by corrosion of steel in concrete, poor quality or deteriorated concrete, and damage to prestressed or post-tensioned tendons embedded in concrete. Discusses each method and includes experiences reported in literature sources and in interviews with state and provincial department of transportation inspection and maintenance personnel. 70 pages. SHRP-S-323, $10

Condition Evaluation of Concrete Bridges Relative to Reinforcement Corrosion. Vol. 2: *Method for Measuring the Corrosion Rate of Reinforcing Steel.* Examines parameters that affect corrosion rate measurements and ranks the most important parameters. Laboratory and field studies were performed using three commercially developed corrosion rate devices. 105 pages. SHRP-S-324, $15

Condition Evaluation of Concrete Bridges Relative to Reinforcement Corrosion. Vol. 3: *Method for Evaluating the Condition of Asphalt-Covered Decks.* Investigates the use of short pulse, ground-penetrating radar to nondestructively identify delaminations at the top and bottom reinforcement levels of asphalt-covered concrete bridge decks. Results help to estimate service life, to programme rehabilitation and maintenance activities, and to estimate quantities for rehabilitation contracts. 84 pages. SHRP-S-325, $10

Condition Evaluation of Concrete Bridges Relative to Reinforcement Corrosion. Vol. 4: *Deck Membrane Effectiveness and a Method for Evaluating Membrane Integrity.* Investigates membrane performance and effectiveness to develop a nondestructive test to evaluate in-place membranes. An ultrasonic pulse velocity method was developed. Conclusions indicate that properly installed and maintained preformed membrane systems reduce chloride intrusion. 143 pages. SHRP-S-326, $15

Condition Evaluation of Concrete Bridges Relative to Reinforcement Corrosion. Vol. 5: *Methods for Evaluating the Effectiveness of Penetrating Sealers.* Two methods, an electrical resistance and a water absorption method, were used to evaluate penetrating sealers for Portland cement concrete bridge structures. A survey of highway agencies in the United States and Canada describes the organizations' experience with various penetrating sealants. 59 pages. SHRP-S-327, $10

Condition Evaluation of Concrete Bridges Relative to Reinforcement Corrosion. Vol. 6: *Method for Field Determination of Total Chloride Content.* Evaluates four procedures to measure the chloride content of reinforced concrete in the field. The methods were used on samples from bridges located in different environments. Includes a detailed test procedure. 155 pages. SHRP-S-328, $15

Condition Evaluation of Concrete Bridges Relative to Reinforcement Corrosion. Vol. 7: Method for Field Measurement of Concrete Permeability. Evaluates a prototype surface air flow (SAF) device for the estimation of concrete surface permeability. A portable field device was constructed that obtains readings at one per minute, allowing a large amount of information to be developed at close intervals across a given concrete member. 87 pages. SHRP-S-329, $10

Condition Evaluation of Concrete Bridges Relative to Reinforcement Corrosion. Vol. 8: Procedure Manual. Describes a procedure to assess the condition of concrete bridge components. Integrates the thirteen applicable, current test methods and procedures with methods presented in the previous seven volumes. Emphasis is on deterioration associated with chloride-induced corrosion of reinforcing steel, but all aspects of durability relative to concrete bridge components are addressed. Designed to be tailored to the needs of a highway agency. 124 pages. SHRP-S-330, $10

Condition Evaluation of Concrete Bridges Relative to Reinforcement Corrosion. Vol. 1 through 8. The entire set of eight volumes of this report. SHRP-S-331, $80

A Guide to Determining Optimal Gradation of Concrete Aggregates. Provides a means to determine the optimal gradation of fine and coarse aggregates for use in the concrete mix using a set of tables. The tables are based on a computer model for the theoretical packing of spherical particles which takes into account their size and specific gravity. Use of these tables in conjunction with the American Concrete Institute's ACI Standard Practice 211.1 should help produce a more workable mix and a better consolidated hardened concrete with decreased permeability and improved durability. 200 pages. SHRP-C-334, $15

Techniques for Concrete Removal and Bar Cleaning on Bridge Rehabilitation Projects. Addresses the partial removal of concrete from decks and other parts of bridge structures. Three technologies are identified and studied in detail: pneumatic breakers, milling and hydrodemolition; analysis addresses work characteristics, production, cost and quality of product. 135 pages. SHRP-S-336, $15

Cathodic Protection of Reinforced Concrete Bridge Elements: A State-of-the-Art Report. Describes the evolution of cathodic protection of reinforced concrete bridges and its current state of the art. Discusses how cathodic protection works, and the effectiveness of the techniques, and an extensive history of cathodic protection of reinforced concrete covering all aspects of completed projects. Design and construction details are available for various types of anode systems. Reviews ancillary equipment such as power supplies and monitoring equipment. Addresses research and development needs required to further the development and use of cathodic protection of reinforced concrete structures. 89 pages. SHRP-S-337, $15

Concrete Microstructure: Recommended Revisions to Test Methods. Analyses and evaluates the results of research performed by SHRP for possible modifications to existing standard methods and specifications from the American Society of Testing and Materials (ASTM), the American Concrete Institute (ACI), the American Association of State Highway and Transportation Officials (AASHTO), and the Pennsylvania Department of Transportation. The evaluation criteria are described. Both specific and general recommendations are made. Implications of results of packing for aggregate grading on ASTM C33 are discussed. An extensive appendix contains trilinear packing diagrams. 107 pages. SHRP-C-339, $15

Concrete Microstructure. Documents the investigation of variables which control cement hydration and microstructural development during the mixing, placing and curing of concrete. 179 pages. SHRP-C-340, $15

Alkali–Silica Reactivity: An Overview of Research. Summarizes current knowledge of how alkali–silica reactivity affects concrete. Targets areas for further research. 105 pages. SHRP-C-342, $15

Eliminating or Minimizing Alkali–Silica Reactivity. Describes various studies of alkali–silica reactivity (ASR) as it affects highway structures. Discusses procedures to evaluate material for safe use in concrete, means to mitigate ASR and its adverse effects in existing concrete and various tests to detect ASR. 266 pages. SHRP-C-343, $15

Concrete Bridge Protection and Rehabilitation: Chemical and Physical Techniques—Rapid Concrete Bridge Deck Protection, Repair and Rehabilitation. Presents the rapid methods for the protection, repair and rehabilitation of bridge decks used by state highway agencies. The report is based on a literature review; responses to questionnaires sent to state Departments of Transportation, Canadian provinces, selected turnpike and throughway authorities, technology transfer centres, and material suppliers; and the evaluation of 50 bridge decks located in seven states.

Compares polymer overlays, sealers, high-early-strength hydraulic cement concrete overlays, and patches for their performance characteristics and service life. 110 pages. SHRP-S-344, $15

Synthesis of Current and Projected Concrete Highway Technology. Summarizes results from a literature review in the fields of concrete materials, construction practices, and applications in highway construction technology. Covers current and projected developments in materials systems including cements, aggregates, admixtures, fibres, and sealers. Other topics include: mix proportioning, batching and transport, placement, finishing, and curing; applications focused on repair and reconstruction, full-depth repairs, slab replacement, partial-depth repairs, overlays, and recycling; and quality assurance methods. Includes a history of concrete pavement construction in Europe. 286 pages. SHRP-C-345 $15

Chloride Removal Implementation Guide. Describes equipment and procedures for the electrochemical removal of chloride from reinforced concrete structures. Provides the basic information needed to implement the chloride removal process on field structures. Discusses pretreatment and post-treatment procedures. 45 pages. SHRP-S-347, $10

Technical Alert: Criteria for the Cathodic Protection of Reinforced Concrete Bridge Elements. Presents results and recommendations based on the investigation of control criteria used to measure corrosion rates of steel in a concrete environment. 14 pages. SHRP-S-359, $5

Concrete Bridge Protection, Repair, and Rehabilitation Relative to Reinforcement Corrosion: A Methods Application Manual. Designed to assist state highway personnel decision-making in protecting, repairing, and rehabilitating concrete bridges exposed to chloride laden environments. Economic models enable selection of the most cost-effective method from a menu of protection, repair, and rehabilitation methods. Present methods such as estimated service life, estimated construction price or cost, construction procedures, quality assurance and construction inspection methods, and material performance specification are described with respect to limitations. Compares two mechanized concrete removal methods, milling and hydrodemolition to the traditional method, pneumatic breakers with respect to labour and capital-intensive operations, work characteristics, and quality management and control. 268 pages. SHRP-S-360, $15

Mechanical Behavior of High Performance Concretes. Vol. 1: *Summary Report.* Describes a literature search and review, the development of mixture proportions of three categories of high performance concrete, the laboratory studies and field trials of the concretes, and the laboratory studies of high early strength fibre-reinforced concrete. Points out the need to remove certain limitations in some of the current specifications that prevent the use of high performance concrete, and concludes with a list of future research needs. Includes two technical guides for the production and use of high performance concrete, and two proposed specifications for test methods. 98 pages. SHRP-C-361, $10

Mechanical Behavior of High Performance Concretes. Vol. 2: Production of High Performance Concrete. Details the laboratory development work on producing high performance concrete for highway applications. Twenty-one different mixture proportions were selected from 360 trial batches for in-depth study and evaluation of the mechanical behaviour of the concrete. The objective was to explore the feasibility of developing appropriate mixture proportions for three different categories of high performance concrete with only locally available, conventional constituent materials and normal production and curing procedures. 92 pages. SHRP-C-362, $15

Mechanical Behavior of High Performance Concretes. Vol. 3: Very Early Strength Concrete. Describes the laboratory investigation, field trials and tests to obtain information on the mechanical behaviour of VES concrete. Tests for the hardened concrete include: compression tests for strength and modulus of elasticity; tension tests for tensile strength, flexural strength, and tensile strain capacity; freezing–thawing tests for durability factor; shrinkage tests; rapid chloride permeability tests; tests for AC impedance; and tests for concrete-to-concrete bond. 116 pages. SHRP-C-363, $10

Mechanical Behavior of High Performance Concretes. Vol. 4: High Early Strength Concrete. Documents laboratory investigations of the mechanical behaviour and field trials of high performance concrete for highway applications. Tests of hardened concrete included compression tests for strength and modulus of elasticity, tension tests for tensile strength and flexural modulus, freezing–thawing tests for durability factor, rapid chloride permeability tests, and various tests on concrete bonding. Field experiments conducted in New York, North Carolina, Arkansas, Illinois, and Nebraska represent a variety of environmental and exposure conditions. 179 pages. SHRP-C-364, $15

Mechanical Behavior of High Performance Concretes. Vol. 5: Very High Strength Concrete. Documents laboratory investigations of the mechanical behaviour of high performance concrete for highway applications.

Hardened concrete tests include compression tests for strength and modulus of elasticity, tension tests for tensile strength and flexural modulus, freezing–thawing tests for durability factor, shrinkage tests, creep tests, rapid chloride permeability tests, tests for AC impedance, and tests for bond between concrete and steel reinforcement. 101 pages. SHRP-C-365, $10

Mechanical Behavior of High Performance Concretes. Vol. 6: High Early Strength Fiber Reinforced Concrete. Provides an extensive database and a summary of a comprehensive experimental investigation of the fresh state and mechanical properties of high early strength fibre reinforced concrete. Fresh properties tested include air content, workability (by the inverted slump test), temperature, and plastic unit weight. Tests on the mechanical properties include compressive strength, elastic modulus, flexural strength, splitting tensile strength, and fatigue life. Sixteen different combinations of parameters were investigated; the variables were the volume fraction of fibres (1 and 2%), the type of fibre (steel, polypropylene), the fibre length or aspect ratio, and the addition of latex or silica fume to the mix. Identifies optimal mixes that satisfied the minimum compressive strength criterion, and showed excellent values of modulus of rupture, toughness indices in bending, and fatigue life in the cracked state. 297 pages. SHRP-C-366, $15

Cathodic Protection of Concrete Bridges: A Manual of Practice. Provides explanation, guidance, and direction concerning cathodic protection of concrete bridge elements to the highway engineer. The manual is divided into three sections: Design, Construction, and Operation and Maintenance. Provides standard specifications for cathodic protection systems for both decks and substructures. Specifications are modelled after those currently being developed under AASHTO, Associated General Contractors (AGC) and American Road and Transportation Builders' Association (ARTBA) Task Force 29 and provide a basis for other sections of the manual. A useful reference for engineers who design and prepare specifications or who oversee turnkey operations. 211 pages. SHRP-S-372, $15

Optimization of Highway Concrete Technology. Summarizes state of the art technology, evaluates test methods for in-place concrete density, and offers guidelines to avoid thermal effects in concrete pavement slab placements and packing-based aggregate proportioning of concrete mixtures. The report also contains descriptions of field evaluations of a variety of concrete mixes used for early opening of full depth concrete pavement repairs and bridge deck overlays, and descriptions of the methodology and content of HWYCON expert system and audiovisual implementation packages for highway personnel dealing with materials testing and pavement and bridge rehabilitation. SHRP-C-373, $15

Field Manual for Maturity and Pullout Testing on Highway Structures. Provides guidance on the use of maturity testing and pullout testing on highway construction projects. Describes background on the use of these procedures together with advice on the selection, and the correct and safe use of testing equipment. Site testing and correlation with standard cured cylinders are described. Gives guidance on the use of ACI and ASTM documents, and a list of recommended publications. 78 pages. SHRP-C-376, $15

Users Guide to the Highway Concrete (HWYCON) Expert System. HWYCON is a software program designed to assist the state highway departments with: diagnosing distresses in highway pavements and structures; selecting materials for construction and reconstruction; and obtaining recommendations on materials and procedures for repair and rehabilitation methods. The package contains seven 3.5 inch high density computer disks and the user's manual. The user's manual provides information about installing the program, the knowledge base on which the program is based, and the operation of the program. SHRP-C-406, $35

Sprayed Zinc Galvanic Anodes for Concrete Marine Bridge Substructures. Steel reinforced marine substructure bridge components are subject to corrosion of the reinforcement caused by chloride contamination of the concrete. This report describes a low-cost method for galvanic cathodic protection by exposing the reinforcing steel, then arc spraying zinc onto the steel and the surrounding concrete to create a 0.5 mm thick galvanic anode. This anode is in electronic contact with the steel and electrolytic contact with the water in the concrete pores.

Field tests in the Florida Keys showed that the anodes retained physical integrity for at least 4.5 years. Laboratory test indicated that concrete resistivity does not represent a main limiting factor in performance of such anodes and that periodic water contact (as encountered in the splash/evaporation zone of marine bridge substructures) is actually necessary for long-term anode performance. This low-cost method is a competitive alternative to impressed current cathodic protection systems and a significant improvement over gunite repairs. SHRP-S-405, $10

Appendix C

Strategic Highway Research Program: unpublished reports on concrete and structures (concrete and the corrosion of steel in atmospherically exposed reinforced concrete bridge components suffering from chloride induced corrosion)

Unpublished reports have limited circulation but are available from the Transportation Research Board (National Academy of Sciences) Publications Department, Box 289, Washington, DC 20055, USA.

The Strategic Highway Research Program (SHRP) was a five-year program that started in 1987. It was administered as a unit of the National Research Council. SHRP conducted accelerated research on a short list of priority research topics: asphalt, concrete and structures, highway operations and pavement performance. The following reports and videos are interim reports or reports expected to be of interest principally to researchers rather than practitioners; thus, they have not been published.

CONCRETE AND STRUCTURES: UNPUBLISHED REPORTS

NCSU Concrete Materials Database. This program was designed to collect and organize research data on the mechanical properties of high performance concrete. Based on the relational model and developed using commercial software, it contains a menu system and other user interfaces that guide users with little database knowledge to extract desired data for data analysis. The NCSU database is the first attempt to demonstrate the feasibility of establishing a general database that encompasses all aspects of concrete properties. See also *User Manual: NCSU Concrete Materials Database Program* (SHRP-C/UWP-91-502). 39 pages. SHRP-C/UWP-91-501

User Manual: NCSU Concrete Materials Database Program. This companion document to *NCSU Concrete Materials Database* (SHRP-C/UWP-91-501), contains instructions on how to log onto the database, query the database for information, and input data. A reference section provides information on hardware and software requirements, as well as further information on querying and adding information to the database. 71 pages. SHRP-C/UWP-91-502

An Electrochemical Method for Detecting Ongoing Corrosion of Steel in a Concrete Structure with CP Applied. Examines the feasibility of using AC impedance spectroscopy (ACIS) as a monitoring tool for detecting corrosion on cathodically protected reinforced steel in concrete. Although the feasibility of the technique was demonstrated for the concrete blocks containing a single reinforcing steel specimen, difficulties in interpretation of the data were created by the large macrocell couples that were present in the large-scale slab tests. 47 pages. SHRP-ID/UFR-91-512

Feasibility Studies on Nondestructive Incorporation of a Conducting Polymer Anode Bed into Bridge Deck Concrete. This feasibility study on the nondestructive incorporation of a conducting polymer anode bed into bridge deck concrete proves that incorporating polymers into concrete by both chemical and electrochemical approaches is feasible. However, it appears that the lateral growth of conducting polymers to form an anode bed requires carefully controlled electrochemical oxidation of the monomer in the concrete matrix. 22 pages. SHRP-ID/UFR-91-517

Quantitative and Rapid Measurement of the Air-Void System in Fresh Concrete. Evaluates a new method for determining air void characteristics in fresh concrete. The method uses a laser counting device to evaluate the number and distribution of air voids. The measurements are made on a core extracted from a sample of fresh concrete frozen with liquid nitrogen. 30 pages. SHRP-ID/UFR-91-519

Evaluation of Electrochemical Impedance Techniques for Detecting Corrosion on Rebar in Reinforced Concrete. Examines the applicability of ultralow frequency AC impedance spectroscopy (ULFACIS) for characterizing corrosion of rebar in concrete. The study demonstrates that ULFACIS can be used to locate and characterize corrosion nondestructively in reinforced concrete structures. A primary objective of the study was to establish whether ULFACIS could be used to determine the polarization resistance, and hence the corrosion rate, of the steel rebar. 98 pages. SHRP-ID/UFR-91-524

Electro-Acoustic Technology as a Means to Modify the Properties of Concrete: A Feasibility Study. Investigates the feasibility of applying Battelle's electro-acoustic technology (BEAT) in the impregnation of organic monomer(s) and in the electro-osmotic removal of chloride ions to stop or retard corrosion in reinforced concrete structures. Results show that using BEAT in the impregnation technique will cause the monomer to polymerize *in situ*, and will cost less than conventional polymer impregnation methods. Although the use of BEAT did accelerate the movement of chloride ion in concrete, research shows that water content is a significant determinant of electrical conductivity. 90 pages. SHRP-ID/UFR-91-526

Expert/Knowledge-Based Systems for Cement and Concrete: State-of-the-Art Report. Surveys the expert/knowledge-based systems applications and development methods related to concrete pavements and structures. The report addresses the following subjects: (1) the potential for the application of expert systems for concrete mixture design and diagnostics, repair, and rehabilitation; (2) a description of inference procedures that are best suited for representing the concrete pavement and structure knowledge domain; and (3) recent expert/knowledge-based systems activities. 31 pages. SHRP-C/UWP-91-527

Alkali Aggregate Reactions in Concrete: An Annotated Bibliography, 1939-1991. Contains nearly 1300 citations, from before 1940 to 1991. It includes numerous contributions from international literature in languages other than English, especially Japanese, French, Chinese, and German. 470 pages. SHRP-C/UWP-92-601

Carbon-Fiber Reinforced Concrete. Concludes that the use of short-pitch-based carbon fibres (0.05% of weight of cement, 0.189 vol. % concrete), together with a dispersant, chemical agents and silica fume, in concrete with fine and coarse aggregates resulted in a flexural strength increase of 85%, and a flexural toughness increase of 205%, a compressive strength increase of 22%, and a material price increase of 39%. The minimum carbon fibre content was 0.1 vol. %. The drying shrinkage was decreased by up to 90%. The electrical resistivity was decreased by up to 83%. 80 pages. SHRP-ID/UFR-92-605

Evaluation of Stratlingite-Hydrogarnet (S-HG) Glass Cement as a Quick-Setting Patching Material. Evaluates S-HG cements, a new type of high alumina cements developed by Corning Glass Works, for use in highway and bridge-deck patching applications. Five blends of S-HG cement are tested and two promising blends identified. Data from these preliminary tests suggest that S-HG cements can be developed into an excellent high early strength highway patching material. 44 pages. SHRP-ID/UFR-92-607

Smart Structural Technology for Nondestructive Evaluation of Concrete. Identifies two specific applications for use of tagged particles in construction materials for long-term condition assessment and quality control. These particles were tested in asphalt and Portland cement concrete mixes. 43 pages. SHRP-ID/UFR-92-608

Identification of Chemical Agents for the Control of Alkali-Aggregate Reaction in Concrete. Identifies a number of chemical compounds for their effectiveness in inhibiting alkali–silica reaction in concrete, either as admixtures in the concrete mix or as penetrating agents to stop further progress of the reaction in already affected or damaged concrete. Two compounds, zinc sulphate and aluminium sulphate, were found to be effective admixtures for fresh concrete. Zinc sulphate also appeared to significantly reduce the subsequent expansion of mortar bars, and could be a suitable penetrating agent for arresting alkali–silica reaction in hardened concrete. 60 pages. SHRP-ID/UFR-92-609

A Literature Review of Time-Deterioration Prediction Techniques. Reviews existing deterioration models used to predict corrosion-related deterioration on reinforced concrete bridges. In addition, the model information was developed using condition ratings provided by technician-inspectors performing visual surveys in accordance with the National Bridge Inspection Standards (NBIS). Discussion of the models illuminates equation definitions, research parameters and lifecycle cost analyses. Included is an annotated bibliography. 150 pages. SHRP-C/UFR-92-613

Freeze–Thaw Resistance in Concrete: An Annotated Bibliography. Contains over 550 citations considered relevant to the phenomenon of freezing and thawing of concrete. Detailed abstracts of studies on the mechanism of frost action as well as case histories and laboratory investigations are provided. Work from fields of ceramics, geology, physics, and soil physics was selected for insight into the roles of moisture movement and ice crystal growth in frost heave and cracking of porous solids. Entries are alphabetical by author or agency. There are author and subject indexes. 227 pages. SHRP-C/UFR-92-617

Cathodic Protection of Reinforced Concrete Bridge Components. Describes a two-year investigation of cathodic protection (CP) systems installed on interstate highway bridges in North America. The performance of 287 systems was reviewed through analysis of questionnaire responses and select field investigations. 80 pages. SHRP-C/UWP-92-618

Evaluation of Norcure Process for Electrochemical Chloride Removal. Provides an analysis of the rate and total amount of chloride removed, the corrosive state of the steel before and after the process, the effects on the concrete, and other aspects of the installations. Comparisons are made to slabs used in other SHRP research on electrochemical chloride removal and protection of concrete bridge components. 31 pages. SHRP-C-620

Development of Metallic Coatings for Corrosion Protection of Steel Rebars. Demonstrates the feasibility of applying a silicon-based diffusion coating on steel rebars, wires and fibres in fluidized beds of Si particles. In comparison to fusion-bonded epoxy coatings, or galvanized bars, the silicon coated samples indicate a higher corrosion resistance in aggressive chloride environments. In addition, the less expensive silicon-coated samples resist scratching. 44 pages. SHRP-I-622

Concrete Components Packing Handbook. Data are based upon a computer model of dry packed, monosized particles adapted from the theories developed by Aims and Toufar in 1967. The model has been demonstrated to adequately describe similar dry packing of powders with varying size distributions in terms of the Rosin-Rammler D' coefficient. The model has been successfully applied to the system cement/fine aggregate/coarse aggregate and has modelled CCA, PCA and PADOT recommended concrete formulations. The results theoretically support the location of recommended concrete formulations in a region of ternary particle mixing which possesses the maximum dry packing density. 161 pages. SHRP-C-624

Maturity Model and Curing Technology. Proposes a new method for determining concrete maturity based on kinetic models of cement hydration employing short-term measurements of heat generated during hydration using isothermal calorimetry. The method uses computer interactive maturity system (CMIS) software. The interrelationship of heat generation, maturity and strength development can be used to predict thermal conditions and strength gain in concrete during curing. 86 pages. SHRP-C-625

Development of Transient Permeability Theory and Apparatus for Measurements of Cementitious Materials. A permeability apparatus was designed and constructed that would allow a rapid and accurate measurement of water transport in concrete. The test specimen acts as a permeable membrane between the two pressurized, large volume reservoirs. Very rapid measurements are possible for specimens possessing permeabilities on the order of microdarcy to nanodarcy. However, in order to measure permeabilities below a nanodarcy the problems of establishing pressure equilibration throughout the test specimen becomes more difficult. 30 pages. SHRP-C-627

Concrete Microstructure Porosity and Permeability. A model has been developed that lays the foundation for relating porosity to permeability. This is based on knowledge gained from previous work as well as experimental and theoretical input from the present program. A linear combination of log-normal distribution may be used to define the pore structure. 86 pages. SHRP-C-628

Cement Paste Aggregate Interface Microstructure. Describes research into the nature of the interfacial region in concrete. The interfacial region, considered more porous than the paste itself, could act both as a localized 'weakness' where fractures are initiated, and as an avenue of attack for aggressive chemical agents. Computer simulations demonstrate that it is the efficiency by which particles pack against the aggregate during mixing which influences the nature and strength of the interfacial region which develops over time. 76 pages. SHRP-C-629

Electrochemical Chloride Removal and Protection of Concrete Bridge Components: Laboratory Studies. Investigates the feasibility of electrochemical removal and concurrent protection as a rehabilitation option for concrete bridge structures. Chloride removal process procedures were developed, and the effects of the process on structure concrete integrity and reinforcing steel were studied. 201 pages. SHRP-S-657

Concrete Bridge Protection and Rehabilitation: Chemical and Physical Techniques – Field Validation. Covers the field application and short-term corrosion performance of six trial installations of two inhibitor-modified concrete systems. The installations were applied to both deck and substructure components in a range of environments. Both pre- and post-treatment corrosion assessments were performed to estimate the corrosion performance of inhibitor modified concrete systems, including visual inspections, delamination surveys, cover depth surveys, chloride contamination levels, corrosion potential measurements, and corrosion current measurements. 67 pages. SHRP-S-658

Concrete Microscopy. Concrete microstructure can be evaluated using both thin and polished sections. Methods described in this report were developed as a supplement to ASTM 856 procedures. The use of an epoxy resin containing a fluorescent dye tended to enhance the ability to view porosity and mechanical features such as interface porosity and cracking. Relationships of formulation and microstructure for a series of 19 concrete samples are presented. A less skilled operator can use the epoxy impregnation technique for developmental and forensic purposes, to more easily observe effects of making and formulation on homogeneity, and the relationship of cracking and secondary hydration products in deteriorated concrete, respectively. 106 pages. SHRP-C-662

Concrete Bridge Protection and Rehabilitation: Chemical and Physical Techniques – Price and Cost Information. Provides an essential component in the process of determining lifecycle costs for ranking alternative techniques for concrete bridge protection and rehabilitation. Data from state highway agencies and toll road agencies in all major geographic regions were utilized. Where some new techniques did not have historical data, costs were estimated using classical engineering estimating procedures. 270 pages. SHRP-S-664

Concrete Bridge Protection and Rehabilitation: Chemical and Physical Techniques – Feasibility Studies of New Rehabilitation Techniques. Examines chemical methods for corrosion protection of reinforcing steel in concrete bridges. A broad spectrum of chemicals were evaluated including corrosion inhibitors, chloride scavengers, and polyaphrons. 170 pages. SHRP-S-665

Concrete Bridge Protection and Rehabilitation: Chemical and Physical Techniques – Corrosion Inhibitors and Polymers. Discusses the improvement of existing non-electrochemical methods for protecting and rehabilitating chloride-contaminated concrete with and without concrete removal and the development of new methods. Five corrosion inhibitors were evaluated and service lives were estimated for the two most effective treatments. Asphalt Portland cement concrete composite (APCCC) was designed and evaluated, and compared with hot-mix asphalts and Portland cement concrete for strength properties, resistance to freeze–thaw and resistance to chloride intrusion. 248 pages. SHRP-S-666

Concrete Bridge Protection and Rehabilitation: Chemical and Physical Techniques – Service Life Estimates. Presents definitions of end of service life of reinforced concrete bridge components exposed to chloride laden environments; categorizes corrosive environments; and defines end of functional service life of untreated bridge decks and substructures. Data from 52 bridge decks distributed in different environmental conditions were collected including chloride contents, cover depths, potentials, corrosion current density estimates, and damaged area measurements. Models were developed to estimate and compare service life of untreated and rehabilitated bridge decks with models based on historical data and time-to-rehabilitate models. SHRP-S-668

Electrochemical Chloride Removal and Protection of Concrete Bridge Components – Field Trials. Discusses the results of field validation trials based on laboratory procedures for electrochemical chloride removal completed on a bridge deck, column substructures, and a bridge abutment within North America. 149 pages. SHRP-S-669

Control Criteria and Materials Performance Studies for Cathodic Protection of Reinforced Concrete. Demonstrates through mathematical models the feasibility of improved and simplified control criteria for cathodic protection of concrete structures. The models were developed to establish concentration profiles which develop as a result of cathodic protection current, and to study current distributions which result from geometric factors. 260 pages. SHRP-S-670

New Cathodic Protection Installations. Presents survey information collected for 36 cathodic protection systems in North America in 1991 and 1992. Eight structures using different cathodic protection systems in different environments were selected for detailed monitoring. 120 pages. SHRP-S-671

Fiber-Optic Air Meter. Discusses a three-phase programme to evaluate both acrylate-filled and diamond-tipped fibre optic air metre probes as well as to gather and evaluate test results comparing fibre optic measurements of entrained air in concrete mix to gravimetric and volumetric measuring methods. 64 pages. SHRP-C-677

Concrete Bridge Protection, Repair and Rehabilitation Video. A description of SHRP's research in these areas. January 1992, 5 minutes. Tape No. 6, $10

Index

Alkali silica reactin (ASR) 35, 86, 170
   cathodic protection with 153
   electrochemical chloride removal 159
   realkalisation 163
Anode
   aluminium 113–14
   corrosion reaction 6, 109, 111, 205
   cathodic protection reactions 108, 109
   probe anode 125, 126
   slot 121, 122
   thermal sprayed zinc 111, 114, 128–33
   titanium mesh 123, 124

Bacterial corrosion 10, 12, 25
Binding of chlorides, see Chloride, binding

Comparison of repair techniques 170
Cover
   importance 199, 218
   measurement 38–9
Carbonation
   depth measurement 51
   see also Diffusion
Cathode
   reaction 6, 10, 11, 205
   under cathodic protection 108, 109
Cathodic prevention 183, 212, 213, 214
Chloride
   binding 22, 24, 28, 53, 56
   content/profiles 23, 56, 189
   content and CP current 139, 145
   threshold for corrosion 23–5, 53–4, 57, 109

Chloroaluminates, see Chloride, binding

Deicing
   salt and its alternatives 2, 217
Diffusion
   carbonates 17, 189
   chloride 187–91

Half Cell
   junction potentials 47–8
Hydrogen embrittlement 109, 147, 159, 163

Incipient anode 93–5
Infrared thermography 36–7
Inhibitors (corrosion) 104–5, 170, 180, 202, 212, 217
iR drop 145

Macrocells 12, 13, 25, 73, 142–3
   corrosion rate probes 73–5
Microcells 12, 13

Passive layer 5, 6, 24
Phenolphthalein 18, 19, 163
Pourbaix diagram 110, 111
Pore
   size 17, 19
   water 16, 17
Prestressing
   failures 25, 26

Sprayed concrete
   overlay for cathodic protection 125
Stray current
   corrosion 10
   and cathodic protection 145, 146, 152, 153

Stray current (*contd*)
   and half cell potentials 47
   and continuity of steel 164

Water/cement ratio 199
Waterproofing membranes 99–102,
   175, 201, 202, 208, 209, 210